Peak Oil

Fossil of giant ground sloth (*Eremotherium*) at the Houston Museum of Natural Science. A cheap, plentiful source of energy (calories) for the first humans to arrive in North America (approximately 15,500 years ago), the giant sloths were extinct within a century.

Peak Oil

*Apocalyptic Environmentalism
and Libertarian Political Culture*

MATTHEW SCHNEIDER-MAYERSON

The University of Chicago Press   Chicago and London

MATTHEW SCHNEIDER-MAYERSON is assistant
professor of social sciences (environmental studies) at Yale-NUS
College, Singapore.

The University of Chicago Press, Chicago 60637
The University of Chicago Press, Ltd., London
© 2015 by The University of Chicago
All rights reserved. Published 2015.
Printed in the United States of America

24 23 22 21 20 19 18 17 16 15   1 2 3 4 5

ISBN-13: 978-0-226-28526-9 (cloth)
ISBN-13: 978-0-226-28543-6 (paper)
ISBN-13: 978-0-226-28557-3 (e-book)
DOI: 10.7208/chicago/9780226285573.001.0001

Library of Congress Cataloging-in-Publication Data

Schneider-Mayerson, Matthew, author.
    Peak Oil : apocalyptic environmentalism and libertarian political
culture / Matthew Schneider-Mayerson.
        pages cm
    Includes bibliographical references and index.
    ISBN 978-0-226-28526-9 (hardcover : alk. paper) —
ISBN 978-0-226-28543-6 (pbk. : alk. paper) — ISBN 978-0-226-
28557-3 (e-book)   1. Hubbert peak theory—Political aspects.
2. Hubbert peak theory—Social aspects.   3. Environmentalism.
4. Petroleum industry and trade—Industrial capacity—Forecasting.
I. Title.
    HD9560.5.S345 2015
    333.8'232—dc23

                                                    2014044790

Portions of chapter 4 were published as "From Politics to Prophecy:
Environmental Quiescence and the 'Peak-Oil' Movement"
(*Environmental Politics* 22.5 [2013]: 866–81) and "Disaster Movies
and the 'Peak Oil' Movement: Does Popular Culture Encourage
Eco-Apocolyptic Beliefs in the United States?" (*Journal for the Study of
Religion, Nature, and Culture* 7.3 [2013]: 289–314).

♾ This paper meets the requirements of ANSI/NISO Z39.48-1992
(Permanence of Paper).

If a sudden vast mass movement could be stamped out of the ground, if thousands could be shaken out of the lethargy and defeatism of centuries by a single speech, how could men doubt that great and world-overturning events would soon come to pass? ERIC HOBSBAWM, *PRIMITIVE REBELS* (1965)

While analysts differ on the details, virtually all agree that the people of Western civilization knew what was happening to them but were unable to stop it. Indeed, the most startling aspect of this story is just how much these people knew, yet how little they acted upon what they knew. NAOMI ORESKES AND ERIK M. CONWAY, "THE COLLAPSE OF WESTERN CIVILIZATION" (2013)

Contents

Acknowledgments

For all their differences, petroleum and libertarian political cultures share a remarkable tendency to break down the bonds that constitute the social and political body. This book, the product of countless conversations with colleagues, friends, and peakists who were extremely generous with their time, was written against that attrition. It would not exist without them.

Many thanks, first and foremost, to Elaine Tyler May and Lary May, whose consistent support, insight, and warmth was invaluable; to Daniel J. Philippon, who provided generous, thoughtful, and detailed feedback and advice on this and other projects; to David N. Pellow, who always provokes me to consider new angles; and to Kevin P. Murphy, a model of critical inquiry and kindness. Thanks to the University of Minnesota for its financial support. Thanks to my colleagues in the Department of American Studies, who patiently listened to and commented on different iterations of this project over the years, especially Melissa Williams, Susie Hatmaker, Danny LaChance, Ryan Murphy, Aaron Eddens, and above all Benjamin Wiggins. Thanks to those who provided advice and support beyond the University of Minnesota, especially David Kinkela, Linda Gordon, Joni Adamson, and Martha Minow.

It would have been impossible to complete this book as quickly as I did without the support of the Center for Energy and Environmental Research in the Human Sciences (CENHS) at Rice University. CENHS and Houston were the perfect institutional and geographic locations to complete a book about energy, culture, and politics.

A special thanks goes out to Dominic Boyer for his intrepid and tireless work in establishing a center for the "energy humanities" as well as to Cyrus Mody, Elizabeth Long, Randal Hall, Alexander Regier, John Zammito, Cymene Howe, Derek Woods, and the students in my spring 2014 "Culture, Energy and the Environment" class.

I am very grateful for the assistance and support of my wonderful editor, Douglas Mitchell, and the generosity of the two anonymous readers.

Thank you to all the peakists I interviewed, including Colin Campbell, Kathy McMahon, Richard Heinberg, Gaea Swinford, Richard Kuhnel, Pat Wentworth, Dee Williams, Oily Cassandra, Kris Can, and many others, as well as all the folks who took the time to answer my surveys. Thanks to Kathy McMahon and James Howard Kunstler, who graciously posted the link to one of my surveys on their blogs. While you may disagree with some of my claims, I hope that the picture I paint is not wholly unfamiliar to you.

Thanks to Laura Wexler, David Graeber, and James Scott, whose engaged teaching and scholarship at Yale inspired me to enter academe in the first place. Along the way, I could not have done this without the kindness of Leslie Gross and Corey Brettschneider. Thanks to all the workers at the coffee shops where I completed much of this book, namely, Tao, Bob's Java Hut, Common Roots, the Birchwood Café, and Double Trouble. Thanks to Blue and Trout Fishing in America for sitting on my keyboard to remind me that interspecies fellowship should always be a higher priority than academic labor. You were correct.

Writing a book can be a lonely process, but good friends certainly made it easier. Thanks especially to Nadia Sussman, Ben Wiggins, Melissa Williams, Christiaan Greer, Shantha Susman, Seher Erdogan, Nick Pinto, David Morawski, Marla Zubel, Charmaine Chua, Chelsea Shannon, Lina Dib, Josh Drimmer, and Cameron Leader-Picone for their friendship. Special acknowledgments go to K. C. Harrison, who helped me struggle with this topic and read early chapter drafts, and Cameron, who read many of these chapters and provided advice and all kinds of support throughout.

Most of all, thanks to my family. To my grandpa Harry, who taught me the value of an open mind, lifelong inquiry, and being a mensch. To my sister Anna, who has always been willing to listen to my half-baked ideas. To my mother Liz, my father Hal, and his wife Rebecca for their love and support. And to my mother, who discussed these ideas with me countless times and showed me that a scholar who is motivated by a concern for social justice can do some good in this world.

Thank you all.

Oil and (Anti-)Politics

To most people who are familiar with the term, *peak oil* is merely a shorthand for energy depletion, but, for a sub-culture of Americans in the early twenty-first century, peak oil became an ideology. Between 2004 and 2011, hundreds of thousands of Americans came to believe that impending oil scarcity would lead to the imminent collapse of industrial society and the demise of the United States.[1] The dire, apocalyptic consequences of energy depletion were a fundamental belief that organized the way they thought about their lives, an "ecological identity" that often superseded other categories of identity.[2] For many, "peakism" transformed the way they lived as well: "peakists" changed occupations, purchased land, retrofitted their homes, stockpiled supplies, and even left their partners as a result of their newfound belief system. This book, the first detailed study of the "peak oil movement," explores the politics of peakism and its implications for contemporary social movements, American politics, and environmental issues, including climate change.

What exactly is the theory of peak oil? One popular explanation was provided by the writer/filmmaker "Oily Cassandra" in her 2007 YouTube video "Porn. Peak Oil. Enjoy" (fig. 1). A split screen was employed to attract and then educate male viewers, who made up over three-quarters of all peakists: on the right side, a sexualized Cassandra, wearing a short pink skirt and a revealing top, performed a mock striptease; on the left side, she soberly explained the concept of peak oil:

Peak oil: it's bigger than terrorism, global warming, or genocide. It's the end of your way of life. It's not right-wing or left-wing doomsday theory, it's the truth, and it's right around the corner. . . . What's the big deal? You'll just drive less, right? The problem is, oil isn't just about driving. Oil affects almost every aspect of our lives. Think about it. Ninety percent of our transportation fuel is oil. And, since most of our clothing, furniture, and resources come from overseas, that means oil-fueled ships may not be bringing them. And what about food? Our abundant supply of food comes from an average of fifteen hundred miles away, which means, if you don't live by a farm, no food for you. There won't be much food anyway since oil is used to make fertilizers and pesticides fired by gas-fueled tractors. If the current farming system stopped using petroleum, soon up to two-thirds of the world's population may not make it. Four billion—billion!—people may not survive.

Even if a Manhattan Project–scale endeavor of renewable energy development were to be implemented immediately, peakists like Cassandra claimed, it would not be sufficient since, "as Asia's booming population begins to demand the same lifestyles as Americans enjoy and as global war starts to stabilize the oil distribution system, the demand" for petroleum will soar. While most Americans have an implicit faith in the development of new technologies to provide energy for future generations, peakists were pessimistic about the potential of other fossil fuels and nonrenewable energy sources to replace petroleum: "It's too late for cheery myths about how technology, wind hybrids, biofuels, hydroelectric dams, solar energy, or hydrogen will save us and let our society continue as is."[3]

Cassandra's video is a window into this subculture's socioeconomic

FIGURE 1 "Oily Cassandra" explains the theory of peak oil in a YouTube video with over 300,000 views.

makeup and virtual orientation. Peakists were primarily white (89–91 percent of those surveyed), male (73–84 percent), and educated—nine of ten had at least a college degree, and one of three had a master's or a Ph.D.[4] The peak oil movement was called the "liberal apocalypse"[5] for good reason—most American believers described themselves as "liberal" (29 percent) or "very liberal" (27 percent), while only 7 percent described themselves as "conservative." Given a wider range of political identities to choose from, more respondents described themselves as "anarchist" (7–9 percent) and "socialist" (10–11 percent) than "conservative" (4–6 percent). Not surprisingly, Cassandra's video was most popular with men, but they were chiefly middle-aged men—in my surveys, the average peakist was forty-seven. Although believers could be found around the world—the Association for the Study of Peak Oil had chapters in twenty-three counties—the phenomenon was focused in the United States (70–77 percent of respondents), where believers were exceptionally passionate.

The medium of Cassandra's evangelism was also reflective of the peak oil phenomenon—like her dramatic persona, the peak oil community existed almost entirely on the Internet. Adherents created a rich virtual community through Web sites, blogs, podcasts, YouTube channels, poems, cartoons, video games, online forums, and even virtual peak oil psychotherapy. Although their primary focus was petroleum, the scope of their concerns is more accurately captured by the phrase *peak everything*.[6] Just as petroleum would become scarce, costly, and eventually unavailable, so too would other natural resources: peak water (freshwater availability), peak dirt (topsoil erosion), peak nature (biodiversity loss), peak fish (overfishing), peak metal (precious metal scarcity), peak uranium , peak natural gas, and so on. Peak oil was, then, not so much a singular obsession with oil as a comprehensive re-evaluation of the relationship between humans, the natural resources we consume, and the environment that sustains us. The potentially far-reaching implications of resource depletion have been the subject of scores of popular books and scholarly articles in a wide array of fields, but the subculture and politics of the peak oil movement have been overlooked, partially because of its virtual orientation—believers generally met on the Internet, prepared for the postoil future alone, and rarely sought publicity.

Peakists stressed the need for collective solutions to oil depletion, reflecting the communitarian ideals of the American Left, but most actually prepared for the post-carbon future alone. After becoming "peak oil aware," three of four began stockpiling food, one of three purchased a more energy-efficient car, one of four moved to a smaller or more energy-efficient home, and one of five changed occupation. Although communal responses to the threat of peak oil received significant media attention toward the end of the first decade of the twenty-first century, only one of four respondents had participated in such collective or political action, and many of these respondents had attended only one meeting.[7] In this way, this group was very different from the traditional social movements with which we are familiar, not buttressed as the latter are by membership organizations with explicit political goals and hierarchies. Instead, the peak oil movement was typical of a twenty-first-century movement organized through the Internet, but it might also be called a *phenomenon*, a *community*, or, borrowing the sociologist Manuel de Landa's term, a *social assemblage*—a fluid, temporally contingent, decentralized grouping centered around knowledge and information across spatial boundaries.[8] As such, it was similar to other loose, decentralized groups, such as the Occupy movement. But, unlike Occupy, the peak oil movement was a movement without (traditional) politics—a movement, in some ways, that was specifically antipolitical.

Even though most peakists followed news and current events more closely after becoming peak oil aware, scanning for signs of the expected collapse, most avoided traditional political engagement after their conversion. In fact, they became less likely to vote or attend marches, rallies, or protests.[9] Cassandra, for example, asserted that "truth has a left-wing bias" and protested the US invasion of Iraq in 2003 but claimed in 2007 that there was no place for her concerns in mainstream American politics.[10] Instead of trying to engage in environmental politics around energy issues and climate change, she told her viewers: "All you can do is slow down collapse to try and give all of us enough time to figure out how to best release ourselves from the age of oil to a simpler, more intimate way of life."[11]

Peakists' political resignation and individualism are particularly surprising because this is the very population that we might otherwise expect to become dedicated environmental activists—liberal or leftist, educated, politically aware, deeply concerned about environmental issues, and cognizant of the need for large-scale change. But, instead of pressuring the federal government, they assumed that "our government is unresponsive, in denial, and ineffectual," as one California man in his late thirties put it, and took individual (or local) steps to prepare themselves and their families.[12] Their struggle to form a collective, political response to a perceived environmental crisis was not uncommon but reflects the decline of trust in social institutions and the rise of a certain type of individualism in the last half century and has troubling implications for and parallels in our response (or lack thereof) to a more immediate contemporary carbon crisis, climate change.

Studies of American individualism have a long history that goes back to Alexis de Tocqueville (1835), if not before, but such claims have intensified over the last three decades. One could point, for example, to two of the most influential scholarly works of the last three decades, Robert Bellah et al.'s *Habits of the Heart* (1985) and Robert Putnam's *Bowling Alone* (2000). Relying on hundreds of interviews, Bellah and his coauthors argued that American culture had succumbed to a therapeutically self-centered narcissism that had stripped public life of its vibrancy and political possibilities. "Utilitarian" and "expressive" individualism had infected common conceptions of the self, the family, civic involvement, religion, the nation, and even "freedom," which by the 1980s meant little more than "being left alone by others, not having other people's values, ideas, or styles of life forced upon one," and "being free of arbitrary authority in work, family, and political life."[13]

Writing a decade and a half later, Robert Putnam marshaled an impressive array of evidence from membership rolls of organizations and studies of social participation to identify a long-term decline of "social capital" (a measure of "social connectedness") in the United States from the 1950s to 2000.[14] Although his methodology was widely critiqued, his basic thesis has generally been accepted and echoed by other sociologists.[15] However, many disagreed with his causal explanations for the decline in social capital, which focus on suburban sprawl, television, and the passing of the "greatest generation" of Americans. More recently, Eric Klinenberg charted the impact of the "cult of the individual" on Americans' living situations: whereas in 1950 only 22 percent of Americans lived alone (accounting for 9 percent of all households), by 2008 over 50 percent of American adults were single (28 percent of all households).[16] This book is in dialogue with these authors, but my study of the peak oil movement suggests that there is a largely ignored political dimension to the changes they observed. I argue that the rise of libertarian ideals in the United States—from the 1960s countercultural concept of idea of "doing your own thing" as a political act, continuing with the fiscal libertarianism of conservatives like Ronald Reagan and the Tea Party, and embedded in recent digital technologies—played a key role in this trend.

In this way, I present this subculture as a microcosm of broader changes that scholars have been actively debating for two decades. Although peakists certainly adopted a counterhegemonic system of beliefs, they were not nearly as exceptional in most other respects. Participants were, for the most part, in the demographic mainstream of the United States midway through the first decade of the twenty-first century—primarily white, middle-aged, and middle-class. Few had suffered mental illness or severe depression, and they were actually less likely to subscribe to popular conspiracy theories than were other Americans. Until their ideological transformation, there was very little to distinguish them from their peers. Indeed, for almost every aspect of the peakist subculture that I discuss there is a significant parallel or overlap with more mainstream activities and developments during the same period. As peakists' vision of the future was influenced by popular narratives of regeneration through crisis in fiction and film, so too did millions of Americans flock to the latest Hollywood disaster movie. As peakists drove less, biked more, and installed alternative energy sources in their homes, so too did many Americans. As peakists tried to consume locally and responsibly, so too were "buy local" campaigns and the "locavore" food movement taking off. As peakists learned to garden, farm, and can their own food, so too did many Americans become interested in such Depression-era activities. As peakists began planning for a post-carbon, climate-altered future, so too were real estate developers, insurers, military planners, and oil and gas companies. As peakists concerned themselves with imagining the post-apocalyptic future, so too did millions of Americans eagerly immerse themselves in zombie (and other post-apocalyptic) films, novels, and video games.

Most importantly, just as peakists were deeply skeptical of the potential for government responses to problems such as resource depletion and climate change, so too were more and more Americans expressing a similar skepticism. By aligning my survey questions with national polls conducted regularly over the last thirty years, I am able to present the peak oil movement as representative of broader currents in American political culture. Peakists considered themselves liberal but exhibited a distinctively conservative response to a perceived threat, and this apparent contradiction has broad implications in the study of both American politics and the environmental challenges of the twenty-first century.

The Libertarian Shift: Conservatism beyond Conservatives

Most American historians who have begun to address the events of the last three decades agree that we live in, or in the shadow of, the "Age of

Reagan."[17] Whether one considers the fortieth president to be the nation's greatest chief executive, as a plurality of Americans claimed in a 2011 poll, or merely the charismatic figurehead for the rising conservative movement, Reagan's ideas, rhetoric, and policies have defined the political culture of the United States for the last thirty years.[18] Even as a Democrat occupies the White House, the vital center of American politics has moved so far to the right that, as one journalist noted in 2012, "virtually every major policy currently associated with the Obama Administration," including the Affordable Care Act (a.k.a. Obamacare), had previously been "a Republican idea in good standing."[19]

Since conservatives came to possess something approaching intellectual hegemony by the second term of President George W. Bush, we might begin to turn our attention from the various causes of the rise of conservatism to its effects. As the historian Kim Phillips-Fein asked, "Are we thinking only of the narrative of the [conservative] movement itself—its rise, its obstacles, its victories? Or are we thinking of the broader changes in American politics—a growing uncertainty about the potential of government, a greater faith in the free market, and a deepening sense of anxiety about collective action?"[20] For thirty years, conservative ideas have dominated American political discourse, and ideas have consequences when paired with power. As Sean Wilentz observed, "the impact of the age of Reagan is indicated" most strikingly not by the passion or achievements of movement conservatives but "by the guiding assumptions and possibilities of American politics and government, and the hold they have on public opinion."[21] In this book, I argue that the individualism of peakists is one measure of this shifting terrain.

Conservative principles of individual choice, reliance on the free market, and the need for a smaller state—libertarian ideas—though unevenly enacted by conservative legislators, are now considered default positions in mainstream political discourse. Furthermore, while American social conservatism has not proved particularly influential overseas, libertarian ideas honed in the United States, whether characterized as *market fundamentalism* or *neoliberalism*, have been exported around the world.[22] This book addresses the impact of conservative governance and intellectual influence beyond movement conservatives, on Americans who do not consider themselves conservatives. I ask questions that are central to understanding the impact of conservatism and contemporary politics in the United States: What is the social influence of the conception of society as merely a collection of individuals? How have repeated criticisms of "big government" and the pervasive distrust of government affected the way individuals conceive of government's ability to address major social problems? How will this *libertarian shift*, as I call it, influence our response to the accelerating environmental problems of our time, such as climate change?

Peak Perspectives

The phrase *peak oil* refers specifically to the peak of global oil production, the point at which half the world's petroleum has been extracted and consumed. Peakists proclaimed not that we are running out of oil but that we will soon witness a scarcity of cheap oil. Evaluating this claim requires an elementary understanding of the process of petroleum extraction. In any oil field, the "energy returned on energy invested" (EROI), the margin of energy acquired from a particular energy source, decreases over time. Initially, reservoir pressure pushes the "black gold" out of newly discovered fields (creating the gushers pictured in films like *Giant* [1956] and *There Will Be Blood* [2007]), but, over time, increasingly expensive technology and risky methods (such as water injection, carbon dioxide flooding, and microbial treatments) must be used to access and recover the remaining oil.[23]

The average EROI on conventional petroleum extracted in the United States in the 1930s was 100:1. That is, for every unit of energy that went into discovery and extraction, the well would yield one hundred times that amount. By the 1970s, it was down to 30:1, and the figure was between 18:1 and 11:1 by the first decade of the twenty-first century.[24] (When EROI has reached 1:1, extraction is no longer profitable.) The estimated EROI for unconventional hydrocarbons, such as Alberta's tar sands and oil shale, is even lower, between 5 and 6:1 and 1.5 and 2:1, respectively.[25] The United States reached its own peak of conventional production in 1970, and by 2009 at least fifty of the fifty-four oil-producing nations had passed their respective peaks.[26] While new oil fields are still being discovered, the rate of discovery has significantly decreased over time, and new finds tend to be much smaller than the giants discovered in the 1960s and 1970s and are increasingly offshore (below the sea).[27] However, new technologies (or expensive technological processes now made economical by rising oil prices) allow producers to recover more oil from older fields, forcing analysts to revise their predictions of when the global peak will occur.

Since there is a finite amount of petroleum on the planet, peak oil is not so much a theory as a geologic fact—the question is exactly when it will occur and whether it actually matters. Given the number of variables involved, there is no way to accurately predict the global peak. The exact amount of proven reserves of petroleum is unknown since many oil-producing countries (such as Saudi Arabia) closely guard this information and many governments tend to (deliberately) overestimate their own reserves.[28] Future technological developments that will allow producers to squeeze more petroleum out of each field are also impossible to predict, as is the public's tolerance for the environmental impact of newer methods of production or unconventional sources of petroleum, such as tar sands and oil shale, or other fossil fuels, such as natural gas. Further complicating the situation, the actual global peak can be deduced only in retrospect, some years after it has occurred, by analyzing global production. As one petroleum engineer put it, "peak oil is a moving target" since oil, as a commodity, "is always a function of price and technology."[29] While peakists asserted that their predictions were based on basic geology and their opponents responded by referring to the quasi-magical machinations of the market, historians will recognize that the timing and form of concerns over oil depletion are as dependent on social, political, and economic factors as anything else. In other words, the peak is, in some senses, socially constructed.

That said, there is some precedent for fears of an abrupt decline. The historian Richard H. K. Vietor recounts that the US peak in 1970 was a surprise even to industry observers—as late as 1968 petroleum journals were confidently reporting that American producers had 2.5 million barrels per day of spare capacity.[30] Given these uncertainties, predictions about the inevitable global peak vary widely. A 2011 forecast by the US Energy Information Administration, for example, claimed that the peak could occur as late as 2067,[31] while the United Kingdom Industry Taskforce on Peak Oil and Energy Security reported in late October 2008 that peak oil would likely occur by 2013.[32] In 2009, the widely respected International Energy Administration estimated that it would occur in 2020, while, in 2010, independent researchers from Oxford University argued that production would peak before 2015.[33]

In contrast, most peakists believed that global oil production was already at the plateau of peak production in 2011. For them, the date of the global peak was critical information: afterward, as Cassandra put it, "the world supply will drop and prices will skyrocket."[34] There would be no orderly, peaceful transition away from the current energy regime dominated by petroleum since awareness of impending scarcity will cause panic and "resource wars" over the remaining oil that would siphon precious funds from the development of alternative energy sources.[35] Skeptics responded that the precise date of a global peak is insignificant since markets will spring into action, pushing producers to exploit old fields in new ways and industry to develop alternative energy sources.[36] They also pointed out that this was far from the first prophecy of oil depletion. In 1855, for example, an advertisement for Kier's Rock Oil advised consumers to "hurry, before this wonderful product is depleted from Nature's laboratory."[37] In 1919, the director of the US Bureau of Mines predicted: "Within the next two to five years the oil fields of this country will reach their maximum production, and from that time on we will face an ever-increasing decline."[38] These concerns were renewed in the 1970s, often by mainstream economists—in 1973, for example, *Foreign Affairs* published an article with the descriptive title "The Oil Crisis: This Time the Wolf Is Here."[39]

While I remain agnostic on the timing of the wolf's arrival and the size of its bite, peakists' broadest claim—that human activity and life are dependent on material resources that are not infinite—is opposed (among those who consider such questions) only by a small minority of futurists, economists, science-fiction authors, and other extreme techno-optimists. As Shane Mulligan put it, oil has become "an essential element of human ecology," and, therefore, "the energy resources on which our economic system relies may be as essential to human life, in their way, as water, biological diversity, or an intact ozone layer."[40] Whether it is imminent or two decades away, the threat of energy scarcity should be taken seriously.

However, as we shall see, peakists were concerned with far more than the availability of a single commodity. They openly disputed, in word and deed, the sustainability of fossil fuel–dependent, growth-based capitalism. As such, the peak oil phenomenon cannot be written off as another case of misguided American apocalypticism but must be understood in the context of what most scientists and scholars are coming to understand as the basic ideological, social, and philosophical challenge of our time: the bedrock incompatibility of our high-energy, high-consumption way of life with a planet of fundamentally limited natural resources.[41] Anthropogenic (man-made) climate change can, in this light, be understood not just as the product of carbon emissions but as the limited capacity of natural sinks (namely, oceans, forests, and soil) to safely sequester that carbon. Climate change is the defining trial of the twenty-first century and a central subject of this book, but it constitutes—as I document in the conclusion—merely one example of a broader conflict between human ideology and the biosphere on which our lives depend.

Peak Politics is primarily about the political and environmental implications of the peak oil movement's response to the threat of ecological crisis, but it is also by necessity about oil itself, rightly labeled "the most powerful fuel and versatile substance ever discovered."[42] While peakists were mistaken about the immediacy and gravity of the oil peak, their assertion that petroleum depletion has the potential to cause "the end of our American way of life," as a Virginia peakist put it, was not.[43] Because of its seeming inexhaustibility and price stability throughout the twentieth century, the importance of petroleum (and energy in general) has rarely been reflected in American scholarship and nonfiction writing—like oxygen, oil has been indispensable, ubiquitous, and appreciated only when scarce. When scholars have written about petroleum, they have traditionally focused their energies on the oil industry, petropolitics, environmental disasters, and individual wildcatters.[44] Following the lead of nonfiction authors, they have recently begun turning their attention to the ways that petroleum has influenced and organized life in the United States.[45]

Statistics—such as the fact that Americans use approximately nineteen million barrels of oil every day—often fail to convey the pervasiveness of petroleum.[46] Consider, for example, the physical manufacture of the book that you hold in your hands. For the cover alone, petroleum was used in the process of tree harvesting and transportation of pulp from China and Russia to South Korea, then to New York on a container ship, and then trucked to a paper warehouse in Massachusetts. The raw materials used for plastic lamination and adhesion originated in Asia, and also contain oil products. The ink, which was sent from a factory in Japan, does not contain petroleum (a change from a few decades ago), but petrochemicals are key ingredients in the ink-varnish used for transfer and drying. Its pages were printed, assembled, and bound in Ann Arbor, Michigan, before being trucked to Chicago, Illinois, and then sent (via truck or airplane) to a regional vendor before finally reaching your hands. Petroleum was used in each of the machines used to harvest trees, operate the paper mill, and deliver the pages, ink, and cover to the printer, whether by truck, airplane, or ship.

If you are reading these words digitally, petroleum was used in every step of your computer's long and complicated commodity chain. Precious metals were mined all around the world, transported via trucks and ships (each trip using petroleum) many thousands of miles, manufactured (along with plastics derived from petrochemicals) in factories, stored in other buildings (often heated using heating oil), before the final product was shipped to consumers—each step involving massive amounts of petroleum or products derived from it. One analysis estimates that each computer requires the consumption of over 150 gallons of petroleum before it reaches the consumer.[47] While most observers are more than familiar with petroleum's primary use as the world's most "energy-dense, easily deliverable source of fuel for transportation," less than half of each barrel of crude oil is turned into gasoline.[48] The rest is used to produce items such as asphalt, pharmaceuticals, detergents, synthetic fibers, artificial colors and flavors for food, refrigerants, all kinds of plastics, and fertilizers for large-scale agriculture. In all these ways, "oil runs through the bloodstream of our daily lives," not just our automobiles.[49] So it runs through these pages as well.

Methodology and Chapter Summaries

Scholars, primarily in the United Kingdom, have already examined the theory of peak oil from a number of perspectives.[50] I utilize this secondary literature throughout this book, but most of my claims are based on original research conducted over a period of five years. My mixed methodology included interviews with peak oil leaders and rank-and-file believers; discourse analysis of movement Web sites and online forums; visits to Transition Towns, where peakists attempted to build resilient, sustainable communities in preparation for peak oil and climate change; and literary analysis of peak oil novels and nonfiction works. The evidence for my claims also relies on two large-scale online surveys conducted in 2011. Both were anonymous and used snowball sampling (or chain sampling) to recruit participants, while specific questions ensured that respondents belonged to the target population.[51] The number of responses was evidence of the size of this community: the first survey, conducted in January 2011, gathered over one thousand responses to multiple-choice and open-ended questions within one week from just two links on peak oil Web sites. (Interested readers can consult appendix 1 for the survey questions.) In addition to hard data, these surveys provided a wealth of detailed responses to open-ended questions—my respondents, as informal members of a group generally ignored or caricatured by the media and often marginalized by their friends, coworkers, and family, were often eager to share their thoughts. As a result, this book contains a level of detail about and analysis of the peak oil phenomenon that has not previously been possible. While peakism is still alive and well in some places, *Peak Poli-*

tics is written in the past tense because it analyzes a vibrant social formation that existed from roughly 2005 to 2011 and does not presume to accurately describe (let alone evaluate) continuing manifestations of peakism or related ideologies.

Chapter 1 introduces the range of Americans involved in the peak oil movement, their motivations, their responses to the threat of collapse, and the social consequences of their actions. I use quotes from my surveys and peakist online forums to explain the ideology and shared experiences of peak oil believers: their quasi-religious conversions, dramatic change in thinking, and reordering of their personal cosmologies. As a window into this subculture, I profile two well-known figures, the evangelist James Howard Kunstler and Kathy McMahon. Kunstler was peak oil's Jeremiah, a one-time journalist whose blog, speaking tours, television appearances, and nonfiction books, such as *The Long Emergency* (2005), spread the peak oil gospel.[52] McMahon, the "Peak Shrink," was a psychotherapist who transformed her practice and personal life after learning about peak oil. In her popular Peak Oil Blues blog and subsequent book, *"I Can't Believe You Think That!" Relationship Struggles around Peak Oil, Climate Change and Economic Hard Times* (2011), she addressed the psychological and emotional fallout of peak oil awareness.[53]

McMahon formulated an alternative profile of psychological health, where acute awareness of ecological crisis and preparation for the collapse of global society was not a symptom of emotional instability or mental illness but a sign of lucidity and health. Conversely, she pathologized "normal" responses to environmental problems, from "rigid Cheneyism" to "neoliberal economanic tendencies."[54] Chapter 1 applies this through-the-looking-glass environmental psychological perspective to an experience shared by almost all peak oil believers: social marginalization as a result of discussing and acting on serious environmental issues, such as energy depletion and anthropogenic climate change. Using Barbara Ehrenreich's and Karen Cerulo's analyses of American optimism,[55] I highlight the ways that peakism deviated from the American "dominant social paradigm" of implicit faith in technology and the inexhaustibility of crucial resources.

Chapter 2 documents and contextualizes the phenomenon of peakism in Bush era. Many believers were drawn to the peak oil theory by three major events that placed petroleum squarely in the public eye: America's invasion and occupation of Iraq, which critics claimed was motivated by geopolitical energy concerns ("blood for oil"); the sudden rise of oil and gas prices after a decade of stable, low prices; and the growing awareness and acceptance of anthropogenic climate change. Through analysis of media coverage, Web site and forum membership statistics, and interviews with influential geologists, authors, and lecturers, I trace the diffusion of the concept of peak oil and the formation of a primarily virtual social movement.

I then place peakism in a much broader historical framework by looking at popular beliefs about abundance, scarcity, and "limits"—often centered around energy availability—in the United States over the last century. From post–World War II optimism buttressed by the full-scale exploitation of petroleum for industrial purposes, to the "Malthusian moment" of the late 1960s and the 1970s, to the reassertion of limitless goals and consumption in the 1980s, these patterns of beliefs in abundance and concerns over scarcity are the backdrop against which recent assertions of limits are often understood.

In chapter 3, I argue that the surprising individualism of peakists is evidence of the influence of libertarian ideas on Americans who do not consider themselves conservative. For the most part, peakists prepared alone, stockpiling goods, conserving energy, changing jobs, and purchasing land. They doubted the ability of political institutions to address contemporary ecological crises. This skepticism of the ability of the federal government to manage major societal issues—whether environmental problems, health care, or desegregation—is considered one of the hallmarks of conservatives, not what one might expect from a group composed of self-identified liberals and progressives. Drawing on public polls and scholarship from political scientists and historians, I view the peak oil movement in the context of a transformation in American political culture over the last three decades: the growth of libertarian ideals. As the result of the conservative movement that fully emerged with Ronald Reagan's election in 1980, Americans of all parties and persuasions have come to conceive of themselves and their citizenship in distinctly libertarian terms—individualistic, market based, and privatized—and this has had a significant impact on politics and social life in the United States.

This shift has been accelerated by the use of digital technologies. The peak oil movement, as a primarily virtual phenomenon, demonstrates the tendency (or "social affordance") of the Internet to create a culture of "networked individuals," participants in fleeting virtual communities that exist only online. In contrast to the popular hype of the Internet's impact on politics, I argue that the virtual medium can also serve to depoliticize participants. Although the Internet has often been compared to the crowd, it lacks most of the characteristic experiences of participating in mass actions, such as deindividuation and collective empowerment; in the case of the peak oil movement, participants also tended to underestimate their numbers and, thus, dismiss their potential for political action. I maintain that these effects are partially the product of the personal computer's design by engineers with a tendency toward cyberlibertarianism and the Internet's formation during the ascendancy of neoliberalism.

Chapter 4 situates peakists' sense of impending collapse in the context of the long history of American apocalypticism and highlights the ways that peakist narratives have borrowed and diverted from previous millennial prophecies. From Puritan writings to the postholocaust novels of recent years, Americans have always been drawn to apocalyptic scenarios. I chart the influence of one recent manifestation of this tendency, eco-apocalyptic Hollywood disaster movies from the period 1990–2010, on peakists' conception of environmental change. Using the genre of peak oil fiction, a close reading of James Howard Kunstler's postpetroleum novel *World Made by Hand* (2008) and debates on online forums, I explore the political implications of the apocalyptic predictions of many peak oil believers, connecting the excitement that many believers felt about the post-peak world to their anticapitalist, anti-imperialist aspirations. At a moment when solutions to ecological crises through electoral politics seem unlikely to many Americans, the prophecy of peak oil provided one means of imagining a significantly different world. Adherents saw peak oil as an imminent, transformative event that might put an end to American imperialism, capitalism, and environmental destruction and deliver a superior, more sustainable future. This revolution would be authored not by elected politicians or social movements but by the petroleum-dependent American way of life tripping over its ecological limits. I suggest that certain aspects of this group's political quiescence reflect a similar dynamic among the broader environmentally aware American Left.

In chapter 5, I use survey responses, online forum comments, and peak oil fiction to explore the racial and gender dynamics of the peak oil movement. Why were so many peak oil believers educated white men? I conclude that some white, male peakists feared that gender equality and immigration would erode their own privilege. These fears were aggravated by long-term political and economic changes and the recent economic recession, which led some male peakists to posit (or hope for) surprising developments in the post-carbon future: the reversal of feminism and a return to a mythical pioneer masculinity. With these issues in mind, I explore the gendered and raced individualism and false meritocracy of many recent post-apocalyptic fantasies, from peak oil visions to the zombie survival scenarios, via an analysis of the television show *Revolution* (2012–14).

The conclusion summarizes the lessons of *Peak Politics* and applies them to our contemporary political and environmental situation. I contend that the very real crises that Americans and the planet now face, which include climate change, a globally interconnected economy, and eventual resource depletion (including fossil fuels), requires a intra- and international communitarian engagement that demands a historical break with the long tradition of American individualism and the more recent culture of optimism. The rise of libertarian ideals (whether embodied by conservatives, moderates, or liberals), the product of America's frontier past, a half century of post–World War II wealth and hegemony, and a Reaganite opposition to any discourse of limits (ecological or geopolitical), will prove ill suited for the near future. The sooner we recognize this incongruity, the better off we will be for it.

ONE

The Peak Oil Ideology and Subculture

I look for ways to use less energy and go lighter upon the land. WHITE MALE, LIBERAL, FLORIDA, FIFTY-ONE TO FIFTY-FIVE

It has changed my entire being. I went from a futurist cornucopian to a doomer in about two years. I gave up my car and ride a bike everywhere now. I installed a little solar, and I take steps to try to prepare for the coming squeeze. WHITE MALE, MISSISSIPPI, THIRTY-SIX TO FORTY

Anyone who points out the ridiculousness of our current way of life and its dependence on a fundamentally unsustainable resource is laughed at as crazy or shunned as a "downer." WHITE MALE, DEMOCRAT, FLORIDA, THIRTY-ONE TO THIRTY-FIVE

As this chapter's epigraphs suggest, the theory of peak oil inspired intense conviction as well as action. Believers described their awareness of oil depletion and environmental crisis in terms that were strikingly similar to those used to describe a religious conversion. They recalled the exact moment it occurred, their emotional response, the dramatic change in thinking, and the reordering of their personal cosmology. Long-held normative assumptions about the natural environment, the nation, the future, and ethical conduct were questioned and often swept aside.[1] Many believers found new occupations, purchased land, and severed ties with friends and family. In this

This chapter's epigraphs are taken from respondents 9566752, 9682014, and 9576993, respectively.

chapter, I explore the ideology of peakism, a belief system that began with simple doubts about energy security and led many to a dramatic personal transformation. I use information culled from my surveys to illustrate the beliefs that peakists held in common and then examine the ways that this belief system was expressed in real-world actions and a rich virtual subculture by profiling the motivations, concerns, and contradictions of three representative peak oil believers and two influential authors and bloggers, the peak oil Jeremiah James Howard Kunstler and the Peak Shrink Kathy McMahon. I conclude by differentiating peakism from the dominant social paradigm held by most Americans, one characterized by an implicit expectation of social progress, unlimited economic growth, and technological solutions to future problems, and document some of the social consequences of adopting this radical ecological identity.

Peakism

Peakism is the ideology of peak oil believers.[2] As the historian Eric Foner defined it, *ideology* is a "system of beliefs, values, fears, prejudices, reflexes and commitments—in sum, the social consciousness—of a social group."[3] Belief in peak oil is much more than an isolated opinion about petroleum reserves; rather, it is an entire system of beliefs since it "poses another Answer for Everything competing for the roles against many hundred other Answers," as an Arizona man put it.[4] Specifically, peakism is marked by many of the same characteristics of recognized ideologies, including internal coherence, intellectual abstractness, specificity, sophistication, dogmatism, and affective investment.[5] While *ideology* is a contested term in contemporary scholarship, in this case it appropriately represents the degree to which peakists adopted a package of beliefs and expectations that set them apart from most Americans and the way that these beliefs were enacted in cultural practices.[6] My use of *ideology* does not carry the pejorative connotation of false consciousness, but it does recognize that the adoption of a counterhegemonic ideology often "entails an aggressive alienation from the existing society."[7]

As described in the introduction, *peak oil* refers to the maximum rate of global petroleum production, after which oil prices are expected to rise steeply. Peakists believe that this point is almost on us or has already been passed, while most geologists and oil industry insiders believe that a peak in conventional global petroleum production is years, if not decades, away. Befitting a "people of plenty," most Americans have a tacit (if unexamined) faith in the development of new technologies to provide energy for future generations, but peakists are pessimistic about the potential for other fossil fuels and renewable energy sources to replace petroleum, not only as fuel for transportation, but also in the production of plastics, pharmaceuticals, and the enormous array of petrochemicals our current way of life demands. They held fast to a set of practical doubts about the feasibility of an impending energy transition that constituted articles of faith so central to their ideology that they are worth explaining in some detail.

Natural gas is difficult to transport, subject to its own production limits, and environmentally destructive, and the United States lacks the infrastructure needed to replace gasoline with natural gas as the primary fuel for transportation—most oil pipelines are too porous for natural gas, for example. Coal is an inefficient energy source (compared to petroleum), and the health and environmental consequences of coal mining and combustion are increasingly considered intolerable. As the Fukushima disaster in Japan showed, nuclear meltdowns could render entire countries uninhabitable, and the question of safe nuclear waste disposal has never been answered. While renewables such as solar and wind power will be critical to energy production in a postpetroleum world, they were seen as not reliable or efficient enough to replace fossil fuels on a global scale. Even if a major national project of renewable energy development were to be implemented, peakists believed that it would be simply too little, too late. Resource wars over remaining petroleum (such as the Iraq War) will siphon funds and expertise that could go toward the development of alternative energy sources. As billions of people in the Global South (in countries such as India and Brazil) and less-developed countries (such as China) seek and achieve a first world, high-energy lifestyle—the average American currently uses over four times more energy per year than her Chinese counterpart—the global demand for a finite resource will only grow.[8]

Despite its name, the peak oil ideology is more accurately described as *peak everything*, reflecting the limits-to-growth environmental paradigm that emerged in the United States in the late 1960s.[9] Peakists were concerned with the threat of an ecological collapse of which petroleum is the principal symbol, and most considered peak oil and climate change to be interrelated crises. A young peakist from Nevada argued that energy depletion is not an isolated issue but "part of a larger pattern including climate change, overpopulation, degradation of our physical earth (deforestation, denuding of farming landscapes, etc.)," and a Virginia woman in her late forties agreed, claiming: "[Peak oil is] part of [a] larger environmental crisis. Other resources are peaking or declining as well (fertile topsoil, wild fish, pollution sinks)." A New Yorker identified the problem even more broadly: "how humans interact with the environment overall."[10] Peak oil is, then, not so much an obsession with oil as a form of radical environmentalism that challenges the fundamental relationship between humans and the natural resources we depend on. Although the *peak* in *peak oil* is a reference to the standard graph of petroleum production (see fig. 2a), which approximates a bell curve, it is also a metaphor: whether a plateau or a cliff, a peak is the point after which an inexorable decline begins. The *y*-axis of the graph in figure 2 represents global oil production, but for many believers it could be replaced with something much broader and more subjective, such as the expectation of future material and social progress, societal complexity, or capitalism—or even collective quality of life, happiness, and peace.

For peakists, basic doubts about the future of energy production led to an ideological conversion and personal transformation. The moment that peakists learned about peak oil or "got it" was often revelatory—the gulf between their conception of the future before and the conception of the future after their awakening is so stark that this moment often cleft their lives in two.[11] A Washingtonian said that his "future plans [were now] framed by a peak oil future," for example, and a Minnesotan in his late twenties said that peak oil "[would] influence all" his "future occupation and lifestyle choice[s]." For most others, the change had already become evident in their lives. A Hawaiian man in his late fifties said that a "complex of environmental issues," including peak oil and climate change, had "defined [his] adult and professional life," while a carpenter admitted: "All of life is now seen

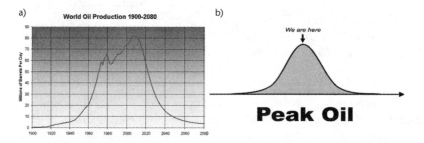

FIGURE 2 *a*, A graph of world oil production to 2004 and prediction of bell-shaped decline. *b*, A popular peakist representation, reproduced on dozens of Web sites and T-shirt designs.

through the lens of a lower energy future, and that has affected all of my choices."[12]

For these Americans, *peak oil* referred to a dramatic teleology in which energy scarcity would lead to the collapse of the late modern capitalist system. Pessimists, sometimes referred to as *doomers*,[13] envisioned an apocalyptic series of events involving warfare over scarce resources, epidemics, famine, and billions of deaths that would leave survivors in a new dark age, while more hopeful adherents imagined a less violent transition into a post-peak world that is simpler, smaller, and more local. There were significant variations in the beliefs of the peakist community, but they were primarily disagreements over the degree and immediacy of the crisis. Some claimed that global oil production peaked between 2000 and 2005 (and is now on a plateau), while others predicted that it was a decade away. Early adherents portrayed the peak oil crisis as a discrete event that would unfold over a few months; others, citing books such as Joseph Tainter's *The Collapse of Complex Societies* (1988) and Jared Diamond's *Collapse* (2005), expected a decades-long decline.[14] Some believed that the effects would not be felt for a few years, while others considered energy prices to be the root cause of the recent economic recession.

Nonetheless, my surveys show that peakists shared a set of beliefs that differentiated them from socioeconomically similar Americans in important ways. While most people who have seriously contemplated these issues (already a minority of Americans) tend to believe that future energy shortages will be solved by technological leaps and renewable energy sources, few peakists agree. On average, they viewed resource wars for petroleum, a halved or quartered global population, and an apocalyptic scenario, including mass epidemics and billions of deaths, as three times more likely than continued business as usual. Almost all expected a significant drop in quality of life for Americans.[15] This ideology is almost the inverse of the expectation of social progress and unlimited economic growth that was still widespread in the pre-recession United States, a contrast I return to at the end of this chapter.

Along with their ideology, most peakists' values and sense of self also underwent a sudden transformation. A Maryland man said, "I am more aware of how my personal lifestyle, as well as that of broader society, is dependent on fossil fuel, and how we will all be affected," while a Coloradan in his late forties was "living more for the moment" as opposed to "planning for the future." A woman in her sixties who saw peak oil and climate change as a "wake-up call that our civilization is going through a paradigm change" said that she had "learned to live with a whole lot less" and was "discovering that what's important to me is not driving an expensive car or living in a five-thousand-square-foot house or flying halfway around the world to go on vacation."[16]

These psychological preparations for collapse and its aftermath were often emotionally painful, and most respondents admitted to short-lived feelings of depression and hopelessness immediately after becoming peak oil aware. However, a surprising number reported that their transformation had been psychologically beneficial. The Mississippi man quoted in one of this chapter's epigraphs claimed that his conversion brought on "the opposite of depression": "It gave me a framework [for my life], and people who believed the same."[17] As a poster to the (now-defunct) Web site "Life After the Oil Crash" put it, many seemed to find "an amazing identity and purpose behind realizing" peak oil: "[It] has helped me take closer notice of life, and spend more time with family and friends, and appreciate the qualities of people and the world round me."[18] A Nevada woman in her early fifties said that she was "living a much simpler life" and "approach[ing] life more a day at a time": "Spending much more time with loved ones and much less time focused on money and possessions. The changes have been good for me and my family." A New York man became "more aware of what [he has] and enjoy[s] 'simple' everyday things like reading a good book or taking a nice brisk walk." A number of believers even found in environmental apocalypticism the motivation to get in shape, such as the Nebraska man who "stays fit so [he] can rely on [his] body in tough times" and the Arkansas man who said: "I slowly begun [sic] to take more care of my physical health because it will be necessary for a day with less energy."[19]

Peak Narratives: From Belief to Personal Action

Becoming peak oil aware constituted an existential shattering of many believers' meaning systems and caused most to change the way that they lived to reflect their revised goals, priorities, and life expectations. A Colorado man in his fifties said that his new "lifestyle is patterned after educating [himself] to how to live in a world with less oil," while a father in his thirties said that his awareness of oil depletion "has changed [his] entire being": "My wife and I are totally restructuring our future." A New Yorker who had helped found the MeetUp group Peak Oil NYC and organized peak oil conferences in 2005 and 2006 said: "[My] awareness of the impending reality of Peak Oil is both directly and indirectly the underlying driver for most of my activities."[20] What actions did he and others take? Nearly three-quarters of American peakists had begun preparing food and supplies for the coming crisis, one in five had changed their occupation to reflect their new interests or require a less carbon-intensive commute, 53 percent drove less, and 32 percent purchased a more energy-efficient car. Only 11 percent had yet to act on their beliefs.[21] As I will show, these actions were primarily personal, not political—for reasons that I explore in chapter 4, only 28 percent of respondents had engaged in formal or traditional political activities related to energy and the environment (campaigns, electoral politics, marches, etc.), and many of these had attended only one meeting of this or that organization.[22] Before exploring the wide range of responses, let us look in closer detail at the peak oil narratives of three representative individuals.

A native Tennessean in his late fifties, "Robert" first heard about peak oil in 2001 and quickly "began reviewing other sources to confirm what [he] had read, as well as paying more attention to current news and analyses about energy topics." He saw oil depletion as a part of "a convergence of problems facing humanity," problems that included "climate change, imperialism, increasing income disparities and poverty, corporatist capitalism" and "infrastructure obsolescence and decay." He described himself as a socialist but was so "disgusted with both parties" that he no longer even voted. After the major peak oil Web sites (described in the next chapter) went online in 2004, he became an active member of the virtual community, checking in multiple times every day until 2010, when he decided to shift from gathering information to active preparation. His life was then organized around the question, "If I prepare now, how can I maintain a good quality of life for my family when energy becomes much more expensive, rationed, or intermittently unavailable?" Despite having never physically met another peak oil believer, he reduced his energy consumption by moving "closer to [his] job and family" as well as "trying to walk more" and "trying to eat less meat"—since (in the United States) meat is two to four times more energy intensive than vegetables and grains—and "trying to buy local food" that had not been shipped from thousands of miles away. He cut down on energy waste by "turning down [the] thermostat in [the] winter, up in [the] summer," and "insulating [his] home." To prepare for the collapse and its aftermath, he was "trying to grow, preserve and store food" as well as "buying bikes, sweaters, long underwear," items that would be in high demand. He also altered his landscape, by "planting trees," as well as his financial strategy, by "investing in precious metals and long-term oil futures." Beyond these tangible preparations, Robert was actively "maintaining awareness" about the timing of the expected crisis and "educating [his] family."[23]

"Mary" from Massachusetts told a similar story of transformation and preparation. She first learned about peak oil in 2005 when she "wanted to become self-sufficient and learn more about gardening and preserving food." She "began perusing Web sites": "I saw the term *peak oil* on Sharon Astyk's site [The Chatelaine's Keys] and the Homesteading Today forum. I picked up a link to The Automatic Earth [a blog discussing issues such as peak oil, climate change, and the economic recession, and] began reading/studying their archives and links to other sites."[24] From 2006 to 2009, she checked these Web sites many times a day. Although she had never suffered from depression before this period, she said that becoming peak oil aware had a significant impact on her mental state—she now foresaw, in the near future, resource wars and a global social collapse, including widespread violence, a die-off of billions of people, and mass epidemics.

Like many believers, she found her new ideology to be socially isolating: "After sharing my concerns with family members and a few friends, I realized very few people were willing to take the issue seriously. They would look at me like I was crazy. When you learn your whole world is going to change, it is overwhelming. Most of the responses that I got were 'we're discovering new oil fields all the time' or 'technology will save us.' My favorite response is, 'If that were true, it would be all over the news!'" Despite never physically meeting another peak oil believer, Mary also decided to transform her life: "We planted fruit trees, grapevines, began gardening and preserving our food. Stockpiled food/seeds. We added chickens and bees to our homestead. Learned to make bread from wheat that we grind. Began assessing energy needs. Considered moving to a smaller place. We've bought some hand tools. We evaluate every purchase in terms of the post-peak world."[25]

"Michael," a fifty-something teacher with a Ph.D., considered himself an anarchist and had been active in environmental politics for many years. After reading an article about the peak oil author James Howard Kunstler in 2006, he "looked up the topic on line and started reading books and articles and following forums like [The Oil Drum]." From 2007 to 2011, he checked these Web sites sites every day because they "provide (mostly) intelligent conversation with (mostly) people who have a clue." Unlike Robert and Mary, he had attended collective meetings, such as Transition Town gatherings and peak oil lectures, and was politically active in the Democratic Party, in the Green Party, and around specific environmental issues. Michael saw oil depletion as an issue endemic to "growth-oriented industrial economies, especially now capitalism with [its] hyperconsumption," exacerbated by "overpopulation." He argued that contemporary capitalism tends to see "the world as something to conquer" and is willing to "burn and break things to get what it wants." As a result, he foresaw an "economic collapse" in the United States, in which the "gov[ernment] [is] unable to fund basic services," with "massive unemployment and mortgage defaults (already [happening])": "Mostly more and more of us falling into the deeply poor with a small super-wealthy class—unless movements overturn them."

Michael noted that climate change was "as much" of a motivation as oil depletion for his numerous life choices: "I don't fly, and mostly don't drive. Picked [a] job close to home. Picked [a] home close to many amenities and good neighbors. Pulled all cash out of stocks before the [economic] crash [of 2008]. Paid off house. Bought an electric car three years ago. Work topics [of peak oil and climate change] into my classes regularly." While emphasizing environmental sustainability in every aspect of his life, Michael agreed that "fictional portrayals of post-apocalyptic scenarios have influenced [his] image of what the post-peak period will be like" and listed the films *Waterworld* (1995), *The Day After Tomorrow* (2004), *War of the Worlds* (2005), *Children of Men* (2006), and *The Happening* (2008) as some of the recent (post)apocalyptic films he has watched. Like most peakists, he found it difficult to bring up the issue of peak oil with others because it is "similar to [climate change] issues—the topic is too big to bring up casually, and most conversation is casual." Despite his immersion in the peakist virtual community, he considered peak oil to be not a "community" or a "movement" but merely a "group of unrelated individuals following their own goals."[26]

In the following chapters, I pick up and explore the cultural, political, and environmental dynamics and implications of the stories told by Robert, Mary, and Michael—the difference between individualistic actions and political or collective engagement, the impact of cultural narratives (such as films) on conceptions of environmental change, and the apolitical tendencies of digital communities—but first I want to document in more detail the kinds of actions that were taken by other respondents. In their responses to contemporary environmental crises, Robert, Mary, and Michael were not exceptions. Most peakists became more actively engaged in planning their own future after becoming peak oil aware, and most said that their chief priority was protecting themselves and their family from the immediate consequences of peak oil rather than educating others or working with others to try to avert the worst consequences. Befitting a twenty-first-century movement, most peakists—61 percent in my first survey, 53 percent in the second—had never physically met another believer. There was no correlation between real-world meetings and real-world actions, which is to say that there were lots of respondents just like Robert and Mary. For example, almost half of the twenty-six respondents to my first survey who had left their spouse or partner as a result of their newfound belief system had never met another peakist.[27] Instead of seeking out physical communities of people with similar beliefs, they sought information and fellowship through the Internet and then transformed their own lives.

Immediate Changes

I divide peakists' responses into two categories: immediate changes and prepping for the post-carbon future. After converting, most believers, as one respondent put it, "look for ways to use less energy and go lighter upon the land," and the easiest and most immediate way to accomplish this goal was simply to drive less. A California man in his early forties said he now "bicycle[s] much more," while a Florida man "started riding [a] bicycle and walking" as a result of his "deep concern for [the futures of his] children and grandchild," cutting his hydrocarbon consumption "from 10 gallons of gas a week to 10 gallons per month on average." Those who resided in cities took advantage of mass transportation, like the "progressive" Washingtonian who started "taking public transit to work," a New York woman who tried "to use public transit more frequently," and a Missouri socialist who "use[d] public transportation exclusively."[28] (In this area, as in many others, peakists were symptomatic of broader shifts as the use of public transportation in the United States increased significantly between 2004 and 2013.)[29] Many peakists were willing to discontinue their leisure pursuits as well, such as the libertarian in his early fifties who "gave up a motorcycling hobby" since it was "for sport and not transportation."[30]

More than half of all Americans are invested in the stock market in some way, and the expectation of imminent collapse led many to adjust their portfolios immediately. Some, expecting a complete collapse, pulled out of the market entirely, such as the Colorado man in his late twenties who "bought gold" and "exited the stock market, excluding some upstream oil and gas investments."[31] Some consulted the peakist investment guides *The Oil Factor: Protect Yourself and Profit from the Coming Energy Crisis* (2005) and *The Coming Economic Collapse: How You Can Thrive When Oil Costs $200 a Barrel* (2006), which predicted an imminent bear market and suggested investments in oil futures and energy companies.[32] A father of two in New York state began to "invest (in every sense of the word) in post-peak assets, from shovels, rakes, and wheelbarrows to firewood harvesting and processing equipment to Schlumberger [a large oil field services company] and Petrobank Energy [a Canadian oil exploration and development company]," and a North Carolina doctor began a "radical change in investment strategies (anticipate economic shocks, invest in oil, gas, gold)."[33] A number of respondents noted that these strategies had already paid off owing in part to the economic recession of 2008.[34] For example, a Minnesota man who "randomly encountered a peak oil Web site" in 2005 said that he developed "a framework for planning ahead that has already yielded financial advantages over what [he] would have done without knowing about the oil peak."[35] The decision to invest in oil and gas stocks is one of many areas where some peakists and mainstream environmentalists, some of whom have organized campaigns demanding that institutions disinvest from fossil fuels, part ways.[36]

Peakists' relationship to a different form of capital, debt, often changed as well. Expecting the system of global capitalism to collapse —and also motivated, no doubt, by the massive debt of many American homeowners after the housing bubble burst—a Virginia woman "focused on paying off loans," while a man in his late thirties took a "second job" to "get out of debt" and "save more money" so as to "purchase precious metals" that would retain their value after capitalism's demise.[37] Others took the opposite approach, like the Colorado man who was so convinced that "the system will . . . collapse or reset based on lack of growth and high unemployment" that he was "not concerned with paying of any debts, especially real estate debt."[38]

Most peakists immediately altered their consumption patterns as well. Acutely aware that national chains have steadily replaced local stores, which could leave them with few options should the global distribution system based on inexpensive petroleum break down, many followed the model of a North Carolina man who supports "local food sourcing" as much as possible and a Washington woman in her seventies for whom "fostering a local food supply has become a full time occupation" since she became an "urban farmer and proprietor of local food store." A Rhode Island woman in her late sixties "joined

[a] local [Community Supported Agriculture group] and cut way back on supermarket shopping" and also began purchasing "clothing that is functional and durable rather than fashionable." With similar considerations in mind, a former activist in New York City said that he had "acquired simple tools and household items made from strong materials that will last a long, long time instead of plastic crap made in China." Others, well aware of the ecological footprint of consumer goods manufactured with petrochemicals and shipped from far away, simply "buy less stuff in general."[39] (Indeed, "buy local" campaigns and anticonsumerist movements have been increasingly popular over the last decade.)[40]

Energy considerations framed not only long-term decisions but everyday concerns as well. A young man from Nevada said: "Everything that I buy I consider can I get it locally, will it be available in a limited energy economy, and should I consider an alternative?" This was true even for respondents who had yet to take action, such as the Ohio man who confessed, "I make plans and think about what I'll do when the collapse happens," and a Wisconsin doctor who is "keeping [his] eye on the horizon, and forming plans to put in place as needed."[41] For a number of peakists, oil depletion was such a constant, habitual concern that they drove more slowly so as to maximize their fuel efficiency and preserve gasoline.

Prepping for the Post-Carbon Future

After these immediate responses, adherents such as Idaho's Richard Kuhnel (fig. 3) began preparing for a future that would look very different from their present. Most embarked on a course of study about topics they previously knew little about, like the Colorado man whose "lifestyle is [now] patterned after educating [himself] . . . to live in a world with less oil." A California woman was busy "collecting books on topics [she] might need in a world with no power sources but human or animal," while a leftist in his fifties "started reading obsessively about agriculture and simple tech[nology]." A New York City resident had "acquired many books on subjects that would be helpful for local living with fewer resources imported from outside a community—foraging, fixing tools." And an environmental activist in her late sixties promised to "continue to educate [herself] and family members about what to expect and how to prepare" by "collecting a library of books and resources."[42] (A stone's throw from such prepara-

FIGURE 3 The Sandpoint Transition Initiative cofounder Richard Kuhnel working in his garden in Sandpoint, Idaho. Photo: Author.

tory literature were remarkably popular pseudoironic survival guides such as *The Complete Worst-Case Scenario Survival Handbook* and *The Zombie Survival Guide*.)[43]

A global collapse and dissolution of nation-states might require survivors to learn skills that were once common but have largely been forgotten (in the United States) as a result of labor specialization. To this end, a central element of the Transition Town movement is the "Great Reskilling," a collective project encouraging and facilitating the reacquisition of the practical skills and know-how that fewer and fewer members of Western industrial society possess.[44] A young man from New Jersey who had become peak oil aware only a few months before taking my survey said that he was "working on developing actual useful skills (woodworking, organic gardening)" and "ha[d] turned away from mainstream entertainment (television, video games, etc.)," while a "progressive" Oregonian in her early thirties "decided not to go back to school and instead just focus on building a cob house [constructed with straw and clay]" and "learn everything [she] can about growing and preserving food on [her] own." A Native American woman had already "learned new skills" such as "fruit and vegetable gardening, canning, fermenting, dehydrating, storing food," while a middle-aged man in Santa Clara planned to go "back to basics" through the "learning of lost arts," including "candle making, furniture building, etc.": "Anything that would help us learn to rely less on imports and commercially made goods and services." Others planned to hone their skills in trades that they expected the post-peak period to value, such as the teacher who was "leaving teaching to start a new career apprenticing" as "a traditional wooden boat builder." "I hope that I can offer these skills to earn a living," she said, and thereby "become a valuable member of a community somewhere."[45] (As I explore in chapter 5, these efforts dovetailed with similar developments in the postrecession United States, such as the local food movement, the "D.I.Y." or "make" movement, and a renewed interest in Depression-era activities such as gardening and canning.)[46]

When the price of electricity is low, as it has been since the 1950s, energy efficiency and distributed (local) power generation are not vital concerns for most Americans. But peakists who owned their homes took on impressive construction projects that they expected would pay off when energy is at a premium. Most tried to increase their efficiency and install renewable home power generation to maximize their self-sufficiency for that day when the electricity grid becomes unreliable. A Nebraska man, for example, had "purchased solar panels" and found "alternative ways to heat [his] house (mainly wood stoves and fireplace enhancements)"; he also "better insulated [his] house, turned down the thermostat, wear[s] sweaters inside, and have heavier blankets." A "liberal" man in his thirties "added solar photovoltaic and solar thermal systems to [his] home," while an architect from Colorado said he had "built a few passive solar houses, one of which [he] live[s] in."[47]

For those who did not already own a home, acquiring land for a post-peak homestead became a priority. With this in mind, a New York woman in her thirties "bought an urban greenhouse with vacant land for cultivation," an African American man in Virginia "purchased a 190-acre farm and started a winery" and was "working to transition there full-time, with the goal of being as self-sufficient as possible," and a Pennsylvania man "bought property" with the intention of "moving off grid." A ham radio enthusiast had "moved out to [his] land in the country" in Alabama, where he "added insulation to the house."[48]

Peakists expect that some regions and cities will be better positioned to survive the twin crises of global warming and peak oil, given their climate and proximity to agriculture, and some made decisions about relocation based on these calculations. When "gas prices spike[d] after Katrina," a Virginia woman began to "reconsider the status quo" and eventually "moved from one state to another [because] of better climate and less dependence on automobile [sic]," while a believer in Queens (New York City) was "moving to Vermont." Similar considerations motivated a man in his early fifties to relocate to rural New York: "I moved six years ago, researched cities with the best water quality and furthest from nuclear reactors and not on the ocean. I now own a very efficient organic farm, and have hybrid vehicles, soon to be totally electric."[49] (While in 2015 most Americans still ignore these risks, urban planners, real estate developers, and insurers now include climate change and related threats in their business models.)[50]

Cheap oil enabled suburban spatial development in the post–World War II period, and gasoline was still inexpensive enough in 2011 (despite an increase of over 300 percent since 1998) for most Americans to live comfortably in the suburbs.[51] Peakists believed that this would soon cease to be the case and that American life would become local once again. A North Carolina man in his late twenties reported that his "family moved to a more traditional 'walkable community,' where [they] can walk or ride bikes to local stores, parks, etc.," while a California man in his late sixties who had "been engaged in social activism since [his] college days (e.g., Vietnam War)" was "actively planning to migrate to [a] small town" where there would be no "need for long commutes by car" and the "impacts of climate change might be less." Other respondents relocated for fear of losing touch with their loved ones when air travel becomes prohibitively expensive. One anarchist, for example, related: "At the time I became peak oil aware, I was living thousands of miles away from where I grew up and where most of my family was residing. . . . I sold my house (sale finalized about 20 months after I became peak oil aware) and moved back to my native state to be nearer to family and friends I had grown up with." After researching peak everything, a scientist in his forties decided to "move from California back to France to be in a more stable place, with better public transportation."[52]

The vast majority of respondents did not move, but they tried to develop as much self-sufficiency as possible. For many, this included engaging in practices that they associated with right-wing survivalists, such as stockpiling goods and even arming themselves. Most peakists, like a Virginia lawyer, were quick to note that they were not "survivalists" since this "implies the hermit in the woods," whereas "an appropriate response" to peak oil and climate change is "simply a return to local agriculture, work/home proximity, closer communities [without a] dependence on things distant."[53] Nonetheless, 72 percent of my respondents stockpiled supplies. The Federal Emergency Management Agency recommends that all citizens store at least three days of canned food and water, but many peakists prepared for a crisis without end in sight. A woman in her sixties reported having stockpiled "food, fuel, survival necessities and tools," while a Georgia man began "stockpiling food and water" after reading James Howard Kunstler's *The Long Emergency*.[54] A Pennsylvanian became so concerned about the future that in 2010 he began "trying to stockpile textbooks so our history and knowledge isn't lost" after the collapse. Given the tradition of nonviolence among liberals and leftists, guns and ammunition were a source of concern for many peakists since their pacifism clashed with their desire for personal preparedness. A "liberal" Washington woman, for example, planned "to buy and learn how to use weapons, which we have never before felt the need to do, which in and of itself is hugely depressing." Many liberal preppers refused to purchase weapons as a matter of principle, like the Arkansas man who had "installed [a] wood stove, bought [a] greenhouse," and "dug up [his] back yard and built garden beds" but insisted: "I have no guns and am ready to help neighbors in my small town."[55]

While a few respondents endeavored to enjoy what they expected to be the last gasps of the Age of Oil—one Floridian was "learn[ing] to enjoy hydrocarbons while we have them," and a Latino man in his thirties took "long driving tours of the US because I know the driving era will soon end"—most were attempting to soften the inevitable transition by modeling the good postcollapse life in the here and now. Although many peakists had previously gardened as a hobby, they now expanded their operations. A Maryland woman in her fifties was "improving [her] soil," an Oregon woman was "keeping chickens," and a Georgia woman said she "now grow[s] two-thirds of [her] vegetables." A California man who had attended the first Association for the Study of Peak Oil (ASPO) conference in 2004 had constructed a "self-sufficient homestead," while a former aeronautics engineer was working on improving his farm, "not so much for [himself] . . . but to give [his] daughters the possibility of a bit longer life span after" the coming collapse.[56]

The Peakist Subculture

The vast majority of these actions were pursued individually—changes in driving and consumption patterns, increased energy efficiency in transportation and housing, and gardening or farming. While comparatively few believers made concerted efforts to form real-world communities or engaged in political actions, many focused their peripatetic energies on erecting an impressive virtual community. Online, they collectively constructed an extensive subculture of peakist films, videos, YouTube channels, podcasts, blogs, songs, poems, cartoons, short stories, novels, and nonfiction books. While the potentially wide-reaching implications of oil depletion have been the subject of dozens of recent scholarly articles in fields as diverse as public health, tourism management, and urban studies, the labyrinthine subculture of peakists has been overlooked.[57] The Internet enables the rapid creation of subcultures and subcultural identities, but this facility can obscure the extent to which these identities are psychologically or emotionally meaningful or expressed in the "real" (offline) world.[58] The facility of conducting discourse analysis of Web sites can lead, on the one hand, to exaggerated claims about the significance of online trends, but the physical invisibility of meaningful virtual communities can also lead scholars and commentators to underreport the significance of these phenomena.[59] Some peakists primarily focused on Web sites, blogs, and online forums, but in the twenty-first-century United States these are crucial sites for the formation of identities and the creation of community.

As I discuss in chapter 3, its reliance on the Internet as its communicative medium led the movement to mirror the architecture of the network itself, but even decentralized, leaderless groups tend to select influential tastemakers. Well-known authors, bloggers, and Web site moderators played a crucial role in popularizing the peakist ideology and shaping its vision of the future. They had much in common with their fans and readers, and their biographies speak to the motivations, concerns, and contradictions of this subculture. As a window into the formation and texture of this primarily virtual community, I introduce two popular and influential figures, the peak oil evangelist James Howard Kunstler and the Peak Shrink Kathy McMahon.

The Evangelist: James Howard Kunstler

James Howard Kunstler was the most widely known and influential peak oil author. In 2007, he appeared at Magers and Quinn bookstore in Minneapolis to promote *The Long Emergency: Surviving the Converging Catastrophes of the Twenty-First Century* (2005). Approximately thirty people sat in the back of the store in gray plastic chairs, muttering excitedly in hushed voices, while another fifteen stood behind them. From snippets of overheard conversations, it was clear that most were already familiar with Kunstler's ideas from his books, blog, or appearances in documentary films. A barrel-chested man in his late fifties with thinning brown hair, a thin moustache, and a mischievous smile, Kunstler was dressed comfortably in blue jeans and a button-down shirt. He was an energetic speaker, pacing back and forth, opinionated and sarcastic. He summarized the argument presented in *The Long Emergency*: as a result of peak oil and climate change, the United States and the world will shortly enter a "long emergency" that will constitute the collapse of the world as we know it, and survivors will experience conditions slightly better than they were in eighteenth-century North America. During the question-and-answer session, a slender man in his twenties stepped forward from the crowd with a hopeful question: "Do you have any advice for young people who will live most of their lives in the long emergency?" Kunstler paused for a moment and responded, without expression: "No." As the crowd filed out, there was the electricity of a brief brush with celebrity and very little evident dejection.

Like many peakists, Kunstler is a baby boomer, born in New York City in 1948. He graduated from the State University of New York at Brockport, where he was a journalist, theater major, and student government president. After college he worked for a number of local and regional newspapers, such as Albany's *Knickerbocker News*, and was a staff writer for *Rolling Stone* in the early 1970s. In 1975 he became a full-time author, publishing seven moderately successful novels between 1979 and 1989, and in the late 1980s he penned a number of articles for the *New York Times* on environmentalism and urban development in the Northeast. His first nonfiction book, *The Geography of Nowhere* (1993), established him as an authority on American suburbia. It presented a critical history of suburbanization, not dissimilar from Kenneth T. Jackson's *Crabgrass Frontier* (1985), alongside a Jane Jacobs (*The Death and Life of Great American Cities*, 1961) model of urban neighborhoods as healthy communities, with special attention to the built environment.[60] *The Geography of Nowhere* was popular enough to be reviewed widely in national media and local newspapers—a typical review, from the *Washington Post*, noted that, although "nothing that he says is really new," Kunstler "has a nice gift for invective and a well-developed sense of outrage."[61] Despite having no formal training in architecture or related fields, Kunstler became known as one of the leading proponents of the architectural design movement New Urbanism, and in 1998 he delivered a keynote speech at the annual meeting of the American Institute of Architects. According to the *Chronicle of Higher Education*, *The Geography of Nowhere* became "standard reading in architecture and urban planning courses."[62]

In early 2001 Kunstler started his own blog, Clusterfuck Nation, and in the wake of the 9/11 attacks he began focusing on energy depletion, oil politics, and then the wars in Afghanistan and Iraq. His career took an unpredictable turn in 2005 when he published *The Long Emergency*, which was read by over 70 percent of my survey respondents. Delivered with Kunstler's sweeping scale and caustic wit, it shared a number of elements with other peak oil nonfiction books: a history of oil production, a description of the centrality of petroleum to our daily lives and its role in modern geopolitics, and an unequivocal dismissal of the potential of alternative energy sources.[63] It stood alone, however, in its far-reaching predictions of the social, economic, and political consequences of the expected crisis. In the long history of apocalyptic literature in the United States, it most closely resembles *The Late, Great Planet Earth*, Hal Lindsey's 1970 best seller, which interpreted the Bible in the context of Cold War geopolitics. In the same way, Kunstler dramatized peakist ideology into an epic narrative of decline, and his vivid imagination struck a chord with many Americans. Thousands of devoted readers echoed the sentiment of a member of the website Peak Oil News and Message Boards who wrote: "I read his book in two or three sittings. I couldn't put it down."[64]

The *long emergency* is Kunstler's term for the drawn-out collapse of the modern world, which had already begun. Like most peakists, Kunstler argued that without cheap and plentiful oil the twenty-first-century world would grind to a halt. American suburbs, dependent on automobility, would become isolated and uninhabitable. Without easy access to local farming cities such as New York and Chicago would be abandoned. Food would have to be grown locally since the cost of transport (via trucks, planes, or ships) would rise, and billions would starve without the petroleum-derived pesticides and fertilizers that enable modern agricultural yields. Competition for remaining resources would cause global wars, and immigrants from poorer nations would overrun their wealthier neighbors. As economies stagnate, jobs disappear, and unrest grows, elites would attempt to hold power by force. Electricity would become scarce. The United States would eventually crumble, with only the Northeast resembling its current condition, while Mexican nationals retake the Southwest, and western states, overly reliant on cars and air conditioning, are abandoned. The Northwest would have more complicated problems since it would "be especially vulnerable to raids emanating from the disintegrating nations of Asia." For those that remain, every aspect of life would be very different, "increasingly and intensely local and smaller in scale." The world would return to a "dark age" as high schools and colleges close and children are put to work. Survivors would see a rise in religious fundamentalism and a return to religious authority; indentured servitude and even slavery may return. Although Kunstler maintained that "just because I say a particular unpleasant thing may happen doesn't mean I want it to happen, or that I endorse its happening," critics accused him of coveting collapse.[65] As a reviewer from Fort Wayne's *Journal-Gazette* put it: "Kunstler takes a curmudgeon's delight in ticking off the many extravagances humanity will have to do without in the years ahead."[66]

His popular blog was cut from this same cloth. Clusterfuck Nation is a freewheeling weekly commentary on American politics, economics, architecture, suburbia, energy, and culture that the *New Yorker* described as "a sustained critique of the cheerful globalism championed by Thomas Friedman."[67] The *New York Times* also took note in 2005, informing its readers that Kunstler had "taken to the Web in a big way" and "moved beyond slash-and-burn suburbia to take on economics and global commerce."[68] The titles of his posts, such as "The Creeping Nausea of American Exceptionalism," "Where Have We Been? Where Are We Going?" and "The First Die-Off," give a sense of his dramatic vision. Despite its dark tone, the site was remarkably popular. In 2008 alone, there were 2.5 million unique visitors, with over 80 percent located in the United States.[69] Clusterfuck Nation had an average of 80,000 visitors a month in 2006, 100,000 in 2007, 175,000 in 2008, and a peak of over 300,000 in September 2008 alone. Almost half of them were categorized as "returning visitors," who might have arrived at www.kunstler.com out of curiosity but returned within the month.[70] Clusterfuck Nation was also syndicated on Peak Oil News as well as Energy Bulletin, a news aggregate site maintained by the Post Carbon Institute.[71]

Kunstler was also busy in the real world. He regularly lectured around the country and appeared in a number of peak oil documentary films, such as *The End of Suburbia* (2005), *Radiant City* (2007), and *Escape from Suburbia* (2007). In each, he played the role of the television scholar to a tee, paraphrasing decades of history into provocative sound bites. His academic reputation made him a frequent spokesperson on peak oil in mainstream news articles, but he was also active within the peakist subculture. For example, he participated in a four-part interview with the performance artist Kris Can on her video blog and YouTube channel "Peak Oil Action and Adventure."

Detour: ASPO's 2009 International Peak Oil Conference, Denver

Can, like Oily Cassandra, attempted to use sex appeal to attract viewers and then educate them about contemporary energy issues. Her signature, sepia-toned video, "Public Service Message," featured a seemingly naked Can swaying behind two oil barrels while delivering a raspy monologue on oil depletion.[72] I learned about her videos at the sixth annual ASPO international conference in Denver in October 2009, where over a thousand people gathered for three days of workshops and panels about peak oil and climate change. ASPO described itself as a "network of scientists and others, having an interest in determining the date and impact of the peak and decline of the world's production of oil and gas."[73] There were ASPO chapters in twenty-three countries, including the United States.

Although his opinionated speculation was anathema to ASPO's scientific, disinterested tone, Kunstler was a regular speaker, and his jeremiads exerted an influence despite his absence that year. Asher Miller, then the director of the Post Carbon Institute ("leading the transition to a more resilient, equitable, and sustainable world"), told me that he first learned about peak oil after picking up *The Long Emergency*; Brian, a thirty-two-year-old electrical engineer who attended the conference because he thought that with rising fuel prices "it might be the last opportunity to fly" anywhere, said his vision of the future was "Kunstleresque."[74] Held in the Sheraton Hotel, the ASPO gathering had the buttoned-down professional atmosphere of a business conference, but a spirit of camaraderie was also evident. Bill Fickas, a forty-seven-year-old libertarian computer programmer who also framed the future in terms of Kunstler's work ("it could go *World Made by Hand*"), said that part of the draw of peakist conferences was the sense of community created by face-to-face meetings: "It helps to find other people who

a)

b)

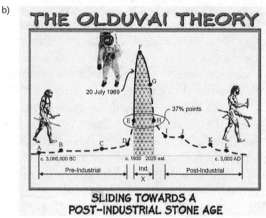

THE OLDUVAI THEORY

20 July 1969

37% points

c. 3,000,000 BC   c. 1930 2025 est.   c. 3,000 AD

Pre-Industrial   Ind.   Post-Industrial
                        X

SLIDING TOWARDS A
POST-INDUSTRIAL STONE AGE

FIGURE 4 *a*, Steve Allen, a participant in the 2009 Association for the Study of Peak Oil international conference at the Sheraton Hotel, in Denver, Colorado. Photo: Author. *b*, A detail of Allen's T-shirt, which combines a graph of global oil production over thousands of years with figures from the classic "march of progress" illustration to predict a "POST-INDUSTRIAL STONE AGE."

share the opinion, and it's a positive thing coming to things like this. Just being able to see that there are other people out there on the Internet is great, but to actually be in a room that if you talked about this they don't think that you're a nutbar, it certainly helps."[75]

Some attendees literally wore their ideology on their clothes. Steve Allen, for example, sported a T-shirt with a graphic depicting the Malthusian "Olduvai theory" (fig. 4*a*), which holds that industrial civilization will endure for approximately one century (a historical "petroleum interval"), to be followed by a "postindustrial stone age." (Such claims, once considered the stuff of conspiracy theories, are now being seriously entertained by mainstream scholars from the humanities and the natural sciences.)[76] Like members of other subcultures, peakists expressed and advertised their identities through commercially produced and distributed goods. Next to us, Smiley Oil, a conference sponsor, was busy demonstrating its educational peak oil video game, *Energy Worlds*—its logo was sinister (but perhaps appropriate to its referent), a cartoonish drop of black gold with a white Cheshire grin. A young woman sold ASPO mugs alongside shirts that proclaimed "I ♥ Peak Oil," and a much wider variety of items could be found online, including bumper stickers, flags, baby bibs, and thong underwear.[77]

Outside the Sheraton, a local oil company thought that the conference enough of a threat to the status quo that it waged a small-scale public relations battle. Aspect Energy, LLC, hired a group of men to dress in chicken costumes and pass out leaflets outside the hotel (see fig. 5). They began: "My oil industry colleagues and I have been fielding lots of questions recently from Chicken Littles worried that petroleum is running out. After I calmed some of them down, they volunteered to hand out a little bit of additional information to anyone who seemed

FIGURE 5 Steve Andrews, cofounder of ASPO, poses with "Chicken Littles" hired by Aspect Energy. Photo: Doug Hansen.

to be casting one worried eye on the sky."[78] The documents listed three "Overlooked Remarkable Facts" about seemingly unlimited petroleum reserves and reprinted a *New York Times* op-ed piece, "'Peak Oil' Is a Waste of Energy.'"[79] Hopeful conferencegoers interpreted the street theater as a sign that perhaps their message was finally being heard.

The Peak Shrink: Kathy McMahon

Kathy McMahon was doing her best to make sure that the message was being heard. McMahon was the Peak Shrink, a practicing psychotherapist, psychology professor, and author of the blog Peak Oil Blues. She began blogging in May 2006 with a candid declaration that she could no longer square her belief in peak oil with her expected professional perspective: "My actions seemed 'irrational' to the pre–Peak Oil mental health professional in me, and looked 'crazy' to those who were unaware, disinterested, or rejected the concept of [peak oil]."[80] While most sites were engaged in debates about the timing and fallout of the oil peak, Peak Oil Blues carved out its niche by examining the emotional and psychological effects of peak oil and climate change awareness. It was so successful that in 2010 McMahon formed a commercial Web site (The Feisty Life) offering one-on-one counseling sessions online and in 2011 published *"I Can't Believe You Actually Think That!" A Couple's Guide to Finding Common Ground about Peak Oil, Climate Catastrophe, and Economic Hard Times.*[81]

Although Peak Oil Blues was not as popular as Clusterfuck Nation, it was far from an isolated outpost in the peakist digital constellation. Other blogs, such as Powering Down: A Journey of Preparation, Dark Optimism: A Better Future for a Troubled World, and Failing Gracefully: Finding Resilience in Uncertain Times dedicated posts to praising Peak Oil Blues.[82] McMahon appeared on several peakist and survivalist podcasts, such as Extraenvironmentalist, Public Service Broadcasting for the Post Carbon World, and The Radio Ecoshock Show, and was featured on peakist YouTube channels, such as the "Women and Peak Oil" episode of Gasoline Gangsters.[83] Her videos were regularly posted on Peak Oil News and Message Boards, where they provoked lengthy debates. Like Kunstler, McMahon's online readers also turned out to see her speak in person—the theme of her 2010 tour was "How to Stay Sane as the World Goes Crazy."

The majority of Peak Oil Blues posts followed a "Dear Abby" style in which the Peak Shrink answered questions posed by regular readers with pseudonyms such as "Homesteader in Paradise," "Off-Grid Girl," and "Preparing to Persevere." The tone was informal, empathetic, and warm, McMahon more likely to playfully speculate on the difficulty of finding the "ideal peak oil mate" than fulminate against excessive energy consumption.[84] The exact number of her regular readers was not available, but the figures from 2011 alone suggest that it was at least in the tens of thousands.[85] In their letters, readers complained of the depression and even suicidal thoughts that resulted from their awareness of peak oil and climate change. "Grim Newlywed," for example, wrote: "I sometimes have moments of terror when I get an intimation of what things will be like when it actually starts to unravel full throttle. . . . [I]t makes me wonder how I could kill myself, rather than suffer things I can't even imagine."[86] The comfort that readers found in McMahon's Web site is attested to in letters and comments by devoted readers such as "Stray Kitty," who asked in July 2008: "How many times have I searched the web trying to find comfort for my anxiety about the future and preparedness? Then I found this site." Another wrote: "What you are attempting to do here, by allowing me and others to vent our deepest fears, is nothing more than a form of liberation. I can only hope that others are lucky enough to be directed here or like me stumble blindly until they find some kind of light."[87]

The Peak Shrink grew up in Boston, the daughter of a firefighter and the third of five children, and still carried her New England accent. After raising two children of her own and earning a doctorate in clinical psychology from Antioch University, she served as the director of the master's program in counseling psychology at her alma mater. Even as she assumed her online persona, she taught in graduate psychology programs and remained a licensed psychologist specializing in couples therapy. Before "learning about Peak Oil," she said, she "saw the world through the eyes of a middle-class US citizen" who believed that "electricity came from light switches."[88] She described her conversion in the same quasi-religious terms that many peakists used: "Then I learned about Peak Oil. After that, I could no longer see the world in the same way. I realized that psychotherapy, while helpful to people in a 'normal' world, could easily become destructive to those with a PO view of the world. . . . I was trying to come to grips with a future cultural transformation that was to be so dramatic, so overwhelming, it disturbed my equilibrium and challenged my very sense of reality."[89]

While her psychological perspective, professional credentials, and lively writing style were part of the Peak Shrink's appeal to her readers, she also served as a model of healthy awareness, engagement, and action. Unlike some of them, who were "living a double life" by keeping "the big secret" of peak oil from their friends and family or were too "paralyzed with fear" to take action, McMahon was psychologically well adjusted and actively preparing for the future.[90] Judging from her frequent anecdotes, she seemed to have fashioned a rural life for herself that the woebegone survivors in Kunstler's *The Long Emergency* would recognize and envy, tending a large garden, and raising chickens. She and her husband lived in a small town in rural Massachusetts where "people cut hay with a scythe, and build homes without power tools."[91]

Beyond providing a model of peak psychological robustness and advice about discussing the peak with family members, the Peak Shrink formulated an alternative profile of psychological health, one in which concerns about environmental crisis and preparation for the collapse of modern society are not a symptom of emotional instability or mental illness but a sign of lucidity and health.[92] This through-the-looking-glass psychology was elucidated in her most popular post, "Do You Have a Panglossian Disorder? or, Economic and Planetary Collapse: Is It a Therapeutic Issue?" The "Panglossian disorder," named after the delusionally optimistic Dr. Pangloss in Voltaire's *Candide* (1759), was defined as "the neurotic tendency toward extreme optimism in the face of likely cultural and planetary collapse." McMahon pathologized "normal" responses as energy anxieties and divided them into fifteen common subtypes, including "rigid Cheneyism ('The American Way of Life is non-negotiable')," "Neoliberal Economanic Tendencies ('A belief that market forces control all—including geological realities')," and "Nascarian Feature" ("People love their automobiles" so "a solution will have to be found to keep us driving"). The post struck a chord with her readers, who responded by saying, "Beautiful article, simply beautiful," and, "Thank you for writing this. It puts my isolated world in perspective."[93]

Ecological Identity and Social Marginalization

McMahon's alternative psychology is one perspective from which we might view one experience that united almost all survey respondents: a painful social marginalization stemming from their ecological identity. Peakists defined themselves according to their environmental beliefs, but they were also shaped by social and cultural forces, such as the responses to their ideology from coworkers, friends, and fam-

THE PEAK OIL IDEOLOGY AND SUBCULTURE

ily. While most scientists report that we are living in an age of environmental crisis, with runaway climate change, widespread environmental toxification, ocean acidification, deforestation, and an ongoing mass extinction of nonhuman animals,[94] Americans whose concern over environmental issues leads them to radically change their lives in a proportionate response—by going beyond green consumerism to bicycle more, drive less, refuse to fly, move to walkable cities, and alter their homes—are still more likely to be considered "crazies" "survival nuts," "hippies," "whackjobs," "tree-huggers," "kooks," and "crackpots" (to borrow a few terms from my respondents) than responsible citizens or human beings. Peakists contended that not being emotionally and/ or psychologically disturbed by contemporary environmental crises was itself the sign of a kind of mental illness. As a Vermont woman put it: "Who wouldn't be depressed with climate change, mass extinctions, suburban sprawl, factory farms, [genetically modified organisms in our food], etc.? You'd have to be nuts to feel otherwise."[95] With these conflicting psychological perspectives in mind, let us explore some of the common reactions that American peak oil believers elicited from their friends, family, and coworkers so as to further differentiate peakism from normative American thinking in the early twenty-first century.

Given their passion and the gravity of their concerns, two of three peakists found it difficult to discuss their environmental beliefs with other people. The responses that they received from their friends, family, and coworkers explain why. A Rhode Island woman noted that "the topic can engender disbelieve [sic] and hostility," while a New York man said: "I have often encountered verbal abuse if I press the issue (and I do not phrase my explanations about peak oil as an end-of-the-world phenomena). It's like I am attacking their entire way of life and their very personhood." A Colorado woman confessed: "The reactions from people make it hard to discuss. I have an old friend that is now barely speaking to me because I discussed this with her." And a New Hampshire man joked that it was "not difficult to talk about" but "so difficult to watch their eyes glaze over and imaginations go catatonic."[96]

Without hearing the other side of these conversations, we have a limited perspective on them, but one frequently cited explanation for this common reaction was the prevalence of optimism in the United States. As scholars from a range of fields have demonstrated, the United States developed in the twentieth century a powerful culture of optimism, which the historian C. Vann Woodward once called a "national philosophy in America."[97] Social scientists such as Neil D. Weinstein have shown that optimistic biases are much more common than pessimistic biases in the United States, and numerous cross-cultural studies have corroborated Americans' tendency toward optimism.[98]

In *Bright-Sided* (2009), Barbara Ehrenreich observed that "the injunction to be positive is so ubiquitous that it's impossible to identify a single source" but cited Dale Carnegie's descendants in the motivation industry as well as the immense influence of positive psychology.[99] The cultural bias against negative thinking is even written into diagnostic psychology manuals, as the sociologist Karen A. Cerulo has observed: "People who routinely focus on the worst-case scenario, those who cannot seem to sustain any real optimism, will likely be diagnosed with dysthymic disorder (commonly known as depression)."[100] A peakist in her early forties reversed this diagnosis, wondering whether "there is [a] collective delusion on the parts of those living in the US, who believe that nothing bad could ever happen here." A number of European peakists agreed, noting that they did not have the "same denial complex in Europe" about environmental issues, while a Canadian found that "even raising the issue of peak oil disturbs North American [*sic*] happy, happy, eternal optimistic attitude." Similarly, a college professor noted that peak oil and climate change are "difficult topics to discuss at faculty meetings or business meetings where optimism toward the [business-as-usual] future reigns supreme": "[My colleagues] are all delusional, of course. Amazing to witness."[101] This sense of invincibility, which I discuss in the next chapter, has been only partially punctured by America's military failure in Iraq and the recent economic recession.[102]

In regularly raising the serious energy and environmental issues that face humanity, peakists were violating the social and cultural norms of attention, conversation, and discussion that govern the issues we can or should pay attention to, discuss in different social situations, and allow to influence us emotionally.[103] As a Connecticut man in his twenties observed, the prevalence of optimistic biases means that many Americans wear "such deeply rose-colored glasses that they interpret peak oil as 'negative thinking.'"[104] And, as Cerulo noted: "When those fixated on the worst cannot or will not be 'cured,' they will likely be distanced from the broader community, their deviance underscored."[105] Since "bad news is bad manners," as a New York woman put it, dozens of respondents described an experience similar to that of a middle-aged Massachusetts man: "[My friends] either say I'm crazy or just roll their eyes and back away." If they weren't viewed as crazy, most peakists were considered downers by friends and family, leading a California man to conclude that "80 percent of Americans do not want to think about the environment, or anything more controversial than who played well or badly in the last football game," while a "progressive" Minnesotan grumbled: "[The] lack of seriousness is growing. Happy faces are preferred." An Oregonian likewise complained: "Due to the brainwashing and the cheer-leading attitudes which prevails [sic] in the US one really can't speak the truth unless it is pleasant."[106]

Energy Optimism: Cornucopianism and Technological Solutionism

On energy and environmental issues, American optimism is manifested as a hegemonic faith in the "dominant social paradigm," a common sense that has existed since at least the post–World War II period with two basic pillars: cornucopianism and technological solutionism. The dominant social paradigm in the contemporary West, which Richard McNeill Douglas has described as "the working faith of our civilization," is characterized by "abundance and progress, growth and prosperity, faith in science and technology, and commitment to a laissez-faire economy."[107] In a capitalist economy dependent on continual growth (of at least 2 percent of GDP per year), those who embrace the dominant social paradigm—that is, most Americans—implicitly accept that the natural resources we depend on are limitless or can be made limitless through technological developments.[108] Buttressing this expectation of infinite growth is a faith in the development of technologies that will solve future problems so deep and so rarely challenged that many peakists said that discussing it was "almost like talking about religion." A North Carolina woman asserted: "The belief in technological progress is a secular religion, including among the academics with whom I interact." And a Colorado man posited: "One must embrace the religious tenets of ones [sic] time, and the main religious tenet of our time is the religion of technology, that technology will save us in the end." As a "progressive" Colorado man who was now "living relentlessly within his means" put it: "Those who do not embrace this religion are viewed as heretics and are shunned by their community." These tenets are rarely considered rationally, as a septuagenarian noted: "When I [ask my interlocutors] how they think a new technology will hold up beneath the weight of scale—i.e., powering millions upon millions of electric cars . . . they don't seem to have an answer."[109] As I discuss in the next chapter, Americans in the post–World War II period saw such incredible technological developments that the nation gradually adopted what Evgeny Morozov terms *technological solutionism*. This unwieldy but useful term signifies, at the most basic level, the expectation that technology will solve future societal problems, but it also recasts "all complex social situations either as nearly defined problems with definite, computable solutions or as transparent and self-evident processes that can be easily optimized"—as opposed to situations that might require concerted political, social, and/or cultural responses or even sacrifices.[110] Faith in technological solutionism is often supported, at some level, with a dose of American exceptionalism, the expectation that "American ingenuity will come up with something else."[111]

Hostile responses to concerns about resource depletion, climate change, species loss, and technological solutionism evinced the age-old tendency to shun the messenger, but they also seemed to hit too close to home. As a North Carolina man noted: "People do not want to look at the limits to their lifestyle. I have found that pointing out flaws and weaknesses in systems is seen [as] criticism of people personally."[112] Claiming that there are limits to infinite growth not only "challenges [Americans'] assumptions regarding the way the world works" as well as "their future expectations"; it "implicates their way of life as part of the problem," as a young peakist observed. As a result of these responses, many believers no longer communicated with their friends, family, and coworkers about environmental issues at all. For example, a North Carolina woman found that "people get intensely angry even at the suggestion our lifestyles may be in for a change," so she "shut[s] up and never mention[s] it." A Georgia woman said: "At this point, I'm very selective about who I'll attempt to talk to about this and even our economic outlook in the U.S."[113]

In the context of a pervasive culture of optimism, what is perhaps most surprising is that a movement based on concerns about energy depletion appeared at all. How and why did this occur?

TWO

Abundance, Scarcity, and Limits in the Age of Oil

In this chapter I present a short history of the peak oil movement and explore the political, economic, and environmental conditions that precipitated its coalescence. Three developments led to the genesis of a social movement motivated by energy concerns: the US invasion of Iraq in March 2003, sharp increases in the price of oil and gasoline, and growing awareness of anthropogenic climate change. Two additional necessary elements were widespread Internet access and the circulation of questions about oil depletion in traditional mass media. Then I place these concerns about resource depletion and scarcity in a broader historical context by looking at popular beliefs about abundance, scarcity, and limits in the United States over the last century. These patterns of beliefs about abundance and concerns over scarcity are the backdrop against which recent assertions of limits—on energy availability as well as climate change—are often understood.

Brief History of the Peak Oil Movement

By the late 1990s, Colin Campbell had spent four decades as a petroleum geology consultant for major oil companies. He had been "crying wolf" (as the *Wall Street Journal* put it) for over two decades about the threat of oil depletion when, in 1998, he penned an article in *Scientific American*, "The End of Cheap Oil," and published a book,

The Coming Oil Crisis.[1] In 2000, he convened a network of "interested scientists and government officials" that called itself the Association for the Study of Peak Oil (ASPO) and began to advocate the theory of peak oil in interviews.[2] In 2002, ASPO held its first conference, the "International Workshop on Oil Depletion," in Uppsala, Sweden. The next year, the second ASPO meeting in Paris received some press coverage, and ASPO created its own Web site.[3]

As a twenty-first-century social movement, the peak oil movement requires a new kind of history, one that is as attuned to page counts and virtual forums as it is to marches and rallies. Campbell's work inspired Ron Swenson, an American engineer and a former engineering professor with decades of experience in renewable energy development and computer security, to create peak oil's first Web site, The Coming Global Oil Crisis, in 1998. In 2001, two additional Web sites popped up, the technical Oil Analytics and the generally Malthusian Die Off.[4] From the era of the civil rights movement until the middle of the first decade of the current century, television offered the most direct and effective means for social movements to gain publicity and promote their cause, but the peak oil movement operated below this level of publicity, in part because it often lacked a powerful visual element— bell curves and oil barrels do not exactly capture the imagination.[5] Although it coalesced on the Internet, traditional ("old") media— meaning newspapers, magazines, and television—played a major role in alerting their audiences to the theory of peak oil. This reflects what the media studies scholar Henry Jenkins calls a *convergence culture*, which has emerged over the last two decades. In this new terrain, old media place "issues on the national agenda" through their wide distribution, but active consumers, working through grassroots media, blogs, and online forums, "reframe those issues for different publics."[6] The peak oil phenomenon was an exemplar of this dynamic, in which the mere mention of the theory of peak oil, even in a negative light, motivated Americans to explore the subject on the Internet, with consequences that the authors of these reports could not possibly have expected or intended.

The first twenty-first-century article on peak oil in American newspapers ran on the Associated Press Business Wire in 2002 and was picked up by newspapers from Dubuque's *Telegraph-Herald* to Connecticut's *Hartford Courant*. Before pronouncing concerns about oil depletion misguided, "Oil Experts Draw Fire for Warning" explained the concept of peak oil and the existence of institutions and experts (such as Colin Campbell) dedicated to it. It introduced a key figure, Matthew

Simmons, described as "an investment banker who helped advise President Bush's campaign on energy policy."[7] By 2003, petroleum trade journals (such as *Oil and Gas Journal* and *Offshore*) begin publishing arguments for and against predictions of an imminent peak of global production, but few imagined the media momentum the subject would gain in 2004, when a flood of books were published, articles and reviews written, and Web sites launched.[8]

Nonfiction books either converted readers or pushed them to search the Internet for "peak oil," which led them to a host of new Web sites, such as The Energy Bulletin, The Oil Drum, Peak Oil News and Message Boards, and Life After the Oil Crash.[9] These sites might serve different functions. The first two, for example, offered updates on energy-related news from sources around the world and expert opinions. The latter two, with large memberships and freewheeling, mushrooming forums, created the sense of community described in the last chapter. When peakists began to speak to friends and family about their new belief system or attend the MeetUp peak oil groups that were forming around the country, the subject was suddenly newsworthy.[10] The peak oil phenomenon, which an Ontario newspaper referred to as the *peak oil cult* in 2004, suddenly merited attention from national newspapers and magazines, such as the *New York Times*, the *Washington Post*, and *Harper's Monthly*.[11]

In 2003, only a handful of titles relating to energy depletion were available, but, in 2004 and early 2005, the following nonfiction books appeared in national bookstores: *The Last Hours of Ancient Sunlight: The Fate of the World and What We Can Do Before It's Too Late* (2004), *High Noon for Natural Gas* (2004), *Out of Gas: The End of the Age of Oil* (2005), *Crude: The Story of Oil* (2004), *The Coming Oil Crisis: Oil, Jihad, and Destiny* (2004), *Crossing the Rubicon: The Decline of the American Empire at the End of the Age of Oil* (2004), *Beyond Oil: The View from Hubbert's Peak* (2004), *Blood and Oil: The Dangers and Consequences of America's Growing Oil Conspiracy* (2004), and *Powerdown: Options and Actions for a Post-Carbon World* (2004).[12] Many were written by established authors, and their dramatic titles give a sense of their tone—each work intending to pierce America's imperial optimism by proclaiming the limits to economic growth, national power, and personal consumption. They educated a public largely unfamiliar with the subject—according to a 2004 Peak Oil News and Message Boards poll, 80 percent of that site's members had learned about the threat of oil depletion between 2001 and 2004.[13]

Concerns about oil depletion spread quickly. Befitting a twenty-first-century social movement, peakism was enabled by growing access to the Internet in the United States. The percentage of Americans with Internet access increased from 44 percent in 2000 to 66 percent in 2004 (and stood at 74 percent in 2013).[14] In March 2004, the first month that Peak Oil News and Message Boards went online, only twenty-five new members joined, but by the end of the year over two hundred members were signing up each month, with the result that the fledgling Web site already had fifteen hundred people actively participating. With new information and analyses posted dozens of times a day, these Web sites confirmed peakists' developing psychological, emotional, and financial investments in the ideology since every new development could be interpreted to vindicate the theory; aided them in converting their family, friends, and others; and helped them plan their futures.[15] Through them, peakists quickly began to develop a subculture that reinforced their concerns. The film *The End of Suburbia: Oil Depletion and the Collapse of the American Dream* (2004), featuring emerging authorities such as Campbell, Heinberg, Simmons, and Kunstler, was released that same year and reviewed in the *New York Times*.[16] By early 2005, adherents were purchasing and wearing "PeakGear," and a wide range of YouTube videos were being posted. In 2004, when Caryl Johnston announced her self-published novel *After the Crash: An Essay-Novel of the Post-Hydrocarbon Age*, she promoted it as "The First Peak Oil Novel" and expected to find an audience that would understand and respond to this claim. MeetUp groups were formed in urban centers, and New York City's was the subject of another *Harper's* piece in 2006.[17]

From 2005 to 2009, peak oil was a relatively popular topic in mainstream American news. The *Washington Post, USA Today*, the *Los Angeles Times*, and *Time* magazine all ran at least one original analysis of the issue, and the Associated Press distributed four articles to its legion of subscribers.[18] As the movement expanded, the concerns of its dedicated members shifted from intellectual study to personal preparation for the post-peak world, and this was reflected in its media coverage. Although the central argument of the peak oil theory—that cheap oil will disappear sooner rather than later—was still the primary assertion that most journalists were interested in (refuting), there was also attention paid to the "movement" (or "cult") of peak oil. The *New York Times* ran six features about the peak oil phenomenon, with titles like "Duck and Cover, It's the New Survivalism" and "The End Is Near! (YAY!)."[19] On radio and Internet news, where subject matter is often more sensational, peakism's presence was far greater—sources ranging from National Public Radio, Salon, and CNBC to Fox News covered the

	2000	2001	2002	2003	2004	2005	2006	2007	2008	2009
IT!	Florida Recount	9/11	Flag Pins	Bring 'em on. / Iraq	Abu Ghraib	Katrina	Watching TV on Computers	Baby Boom	The Election	The Economy
NEWISH	Tiniest Phone	Airport Security	Guantanamo	Friendster MySpace	Wiretaps	Self-Portraits	Tsunami	Rock Bands	Moms on Facebook	Iran on Twitter
BUSINESS	Rolling Blackouts	Dot-Com Crash	S650 / 1BR, A/C, WIFI, W/D, RR, NO-FEE, COZY / Craigslist	Credit-Default Swaps	GOOG 85.00 / Google I.P.O.	Blackwater	¥ / China	Housing Boom	Foreclosures	Stimulus
FEAR	More Is Not Enough	Anthrax	Snipers	SEVERE HIGH ELEVATED GUARDED LOW / Everything	24 / It's Too Late	I.E.D.'s	H5N1 / Avian Flu	$322 \tfrac{9}{10}$ / Peak Oil	Credit Freeze	H1N1 / Swine Flu
MAVERICK	John McCain	NO / Russ Feingold	Al Jazeera	We're ashamed the president... is from Texas / Dixie Chicks	AIR AMERICA / Al Franken	George Bush doesn't care about black people. / Kanye West	Sincere Beards	T. Boone Pickens	REVOLUT / Ron Paul	$ / Goldman Sachs
CHAMPION	Shaq & Kobe	Dale Earnhardt Sr.	Patriots	Steroids	Mia Hamm	Lance Armstrong	Barbaro	Tiger Woods	Michael Phelps	27 / Yankees, Again
CULTURE	Pokémon	Wikipedia! / *According to Wikipedia	"American Idol"	HEY YA! / OutKast	Camera Phones	M.M.O.R.P.G.'s	yummo! / Rachael Ray	Writers' Strike	The Art Market	Lady Gaga
COUPLE	Carrie & Mr. Big	Harry Potter & Voldemort	€ / Europe & Money	★☆☆☆☆ Gigli / Bennifer	Demi Moore & Ashton Kutcher	Tom Cruise & Oprah Winfrey	I can't quit you. / Heath Ledger & Jake Gyllenhaal	Brangelina & Family	Federer & Nadal	$+8 \div 2$ / Not Jon & Kate
FAD	Going Viral	JONATHAN FRANZEN CORRECTIONS / Oprah's Book Club	Collagen	Tuscany	Brownie & Big Time & Rummy & BoyGenius & Condi. / Nicknames	Movies in the Mail	Ironic Mustaches	Crocs	I'm NOT A Plastic Bag / Canvas Totes	VV / Vampires
LOGO	Lattes	FDNY NYPD / First Responders	BADA BING! / HBO	O / Oprah	W04 / Dubya	News vs. News	Apple	LUKOIL / Russian Moguls	Obama	citi
NOUN	glitch	news cycle	freedom fries	spider hole	friendly fire	truthiness	chatter	surge	hope	Auto-Tune
VERB	I.M.	outsource	download	punk'd	Swift boat	Google	text	blog	go rogue	crowd-source
	2000	2001	2002	2003	2004	2005	2006	2007	2008	2009

FIGURE 6 In Philip Niemeyer's retrospective *New York Times* Op-Ed Chart of December 27, 2009, peak oil is represented as the national "fear" of 2007.

movement numerous times during this period. Indeed, in a 2009 op-art graphic in the *New York Times* (fig. 6), Philip Niemeyer portrayed peak oil as the national "fear" of 2007.[20]

Although evaluations of contemporary media ecology tend to downplay the reach and influence of traditional media, the impact of the mere mention of oil depletion in popular newspapers, in magazines, and on television cannot be understated. In late May 2006, for example, Samantha Gross of the Associated Press wrote an article about peakists—"Energy Fears Looming, New Survivalists Prepare"— that was picked up by news sources across the country, from MSNBC .com and *USA Today* to the popular conservative Web site The Drudge Report.[21] This story was unusual in that it mentioned the URL of a specific Web site, Peak Oil News and Message Boards. On that site the number of new members per month had fluctuated between 250 and 350 users during the previous year, but 2,928 people joined in May 2006 alone.[22]

A similar event occurred in early June 2008, when the same Web site was mentioned on the CNN *Saturday Morning News*. During a segment on oil prices, the CNN correspondent Deborah Feyerick explained: "Peak oil is the point when global oil production peaks then goes down. The remaining supply is limited and will be harder to get at, and that means fewer barrels a day. Some oil experts say that day is here, others predict it is twenty to thirty years away. But as gas prices rise, Web sites like peakoil.com and survivalblog are getting more and more visitors talking about the end of cheap oil and the possible threat of political and economic instability around the world."[23] After this story, tens, if not hundreds, of thousands of Americans visited these Web sites, and almost five thousand became members of Peak Oil News and Message Boards the next month.[24]

Media attention directed Americans to peak oil Web sites, but it did not lead them to become members or alter their lives to prepare for an apocalyptic collapse of civilization. They were pushed and pulled in this direction by three of the central events of the the period 2000–2010: the invasion and occupation of Iraq, rising gas prices, and the spread of apocalyptic environmentalism, driven by concerns about climate change.

America in Iraq: Blood for Oil?

The controversial US invasion and occupation of Iraq in 2003 led many Americans to consider global energy politics for the first time in a generation, if not their lifetimes. Beginning in 2002, when the groundwork for the offensive was being laid, and increasingly as President Bush's official justifications for the war (Iraq's possession of nuclear or biological weapons) failed to materialize in 2003, the media speculated on potential hidden motives for the action. Bush's critics, citing "blood for oil" as the president's secret motivation, mounted the largest group of coordinated protests in the nation's history (in February and March 2003), but the Bush administration's adept public relations strategies and the failure of mainstream American journalism after 9/11 kept these claims on the political fringe.[25]

Antiwar protesters, pacifists, and leftists began to oppose the imminent attack in 2002, their efforts culminating in a day of coordinated global protests on February 15, 2003, in over sixty countries. Over 150 different American cities participated, with somewhere between 100,000 and 300,000 people marching in New York City alone.[26] Although protestors criticized the imminent invasion from a variety of political positions, the slogan "no blood for oil" was a unifying cry. The mainstream media tended to bury this critique and rarely elaborated on its potential insight or veracity.[27] By late 2004, when it became clear that Iraq did not possess chemical or nuclear weapons, the primary justification for the war was simply shunted to the sidelines, and politicians began to trumpet the liberation and democratization of Iraq as their ultimate goal. But few were convinced: in a 2007 poll, 73 percent of Americans said they believed that controlling Iraq's oil supply was at least a factor in the decision to invade Iraq.[28] Once a marginal, conspiratorial whisper, the claim that "the Iraq war [was] largely about oil," as Alan Greenspan put it, became widely accepted.[29]

Peakism provided a ready-to-wear ideology for viewing the Iraq War not as an isolated occupation but as a resource war heralding a post–Cold War era of conflict. Nonfiction books such as Michael Klare's *Resource Wars* (2001) predicted that oil scarcity would lead nations into prolonged resource wars for the remaining petroleum; as the official rationale for the invasion vanished, the Iraq War seemed to fulfill the prophecy. Many peakists discovered the peak oil theory while investigating its government's "true" motives. A North Carolina man, for example, said that "answering the question of why we were in Iraq led inevitably to oil and energy resources," while a woman in her sixties "wondered what in the world could cause the US to preemptively attack a country (Iraq) that was not linked to 9/11," until she "discovered the Colin Campbell projections." A Connecticut Democrat said that "the invasion of Iraq makes little sense until viewed from a position that takes into account resource scarcity," meaning that peakism "provid[ed] a worldview" with "explanations of events and trends that makes more sense than the popular narrative provided by more mainstream media outlets."[30]

Rising Gas and Oil Prices

The second event that drew Americans toward peakism was the rising price of gasoline throughout the first decade of the twenty-first century.[31] The United States developed a unique relationship with oil throughout the twentieth century as it became the lifeblood of the country's rise to superpower status. The energy history of the United States is not the subject of this book, but this backstory was so well-known by American peakists—most introductory works on peak oil included a lengthy section on the subject—that it constitutes a cornerstone of their ideology.[32]

The combination of large, easily accessible oil fields, technological expertise, and industrial infrastructure put the United States in the driver's seat of global production from roughly 1860 to 1940, when it still produced two-thirds of the world's oil.[33] Americans consumed most of it, too: by 1929, 78 percent of the world's cars were in the United States.[34] Petroleum first emerged as a major source of national power during World War I, when the United States supplied over 80 percent of the Allies' oil, fueling decisive military innovations such as submarines, airplanes, and tanks. Access to oil was a crucial factor in the Allies' victory in World War II as well—America's petroadvantage was overwhelming, and energy considerations were a crucial (if often overlooked) factor in various military decision and outcomes, such as Hitler's decision to invade the Soviet Union and the Japanese attack on Pearl Harbor.[35] In both wars, then, the United States "floated to victory upon a wave of oil."[36]

In the post–World War II period, the "century of oil" fueled the "American Century."[37] Cheap oil enabled the rise of the automobile, which became the signature commodity of the postwar years and produced elemental transformations in social life, such as the rapid growth of interstate highways and decentralized suburbs. Cars provided physical mobility, a "psychological mobility that freed the individual from the limits of space and time," and a symbolic social mobility (driving the American Dream).[38] As the historian Lizabeth Cohen observed, the rapid expansion of consumerism in the postwar period relied on petroleum-based products, and cheap oil was responsible for "regularizing overconsumption and making it the new normal."[39] New products were suddenly inexpensive enough for mass consumption as "cheap oil allowed chemists to derive cheap replication of costlier products" and "reliance on these products helped define basic patterns of consumption in twentieth-century America."[40] Cheap oil made for cheap food, too, as "oil-powered machinery and petrochemical-based pesticides, herbicides, and fertilizers also sparked [the] unprecedented increases in agricultural production" known as the Green Revolution.[41] In all these ways, this suddenly cheap, price-stable, abundant resource provided "the building block for a proliferating array of consumer goods" and, thus, "underpinned a steadily rising U.S. standard of living" in the post–World War II period.[42]

The 1973 OPEC embargo brought petroleum to the forefront of national attention for the first time and changed the rosy relationship between Americans and oil, but behind the scenes black gold would soon deliver another windfall to the United States. By the 1980s the Soviet Union quietly became the world's largest oil exporter, its economy and ability to support its satellite states heavily reliant on the price of petroleum. The oil glut of the 1980s, which halved the price of a barrel of crude, devastated the Soviet economy and expedited the collapse of the Soviet Union.[43]

The price of both crude oil and gasoline was relatively steady throughout the 1990s before skyrocketing between 2000 and 2009, during which period the average price of a gallon of gas in the United States increased by over 300 percent while the cost of a barrel of crude increased by almost 600 percent.[44] For crude, this jump reflected not any real scarcity but investors' hopes and fears, the unpredictable outcome of public relations campaigns and rumors. The price of gas rose slightly after 9/11 on general fears of instability in the Middle East, hovering around $1.50 per gallon, before beginning to increase in December 2002. While many commentators pegged this increase to concerns about Venezuela establishing an oil embargo against the United States, the primary cause was the threat of war with Iraq.[45] Even the quick victory that neoconservatives predicted might remove Iraq's oil from the global market for some time, and that might lead to a global shortage. OPEC's promise to increase production in the case of such an event did not convince investors, and fears of scarcity drove the price up. The factors involved in determining the price of gasoline are inscrutable to many consumers, and the peak oil theory provided one commonsensical explanation.

Oil and gas prices rose steadily until 2008, with two major spikes (see fig. 7). The first was the result of the actual invasion of Iraq—the week of "Shock and Awe" the average gallon cost $1.72, a 25 percent increase from only three months before. The second was the result of Hurricane Katrina, which destroyed over forty offshore oil platforms

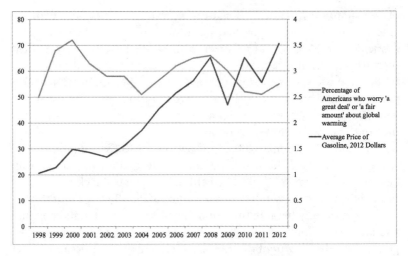

FIGURE 7   The price of gasoline rose throughout from 1998 to 2012, while the percentage of Americans concerned about climate change ebbed and flowed. Sources: Gallup and US Energy Information Administration.

and forced the temporary closure of nine nearby refineries, generating renewed fears of a shortage.[46] Immediately after Katrina, the price of crude jumped almost $7.00 from the previous month. Fluctuations in oil and gas were followed so closely on television, in print, and on the Internet—from December 2002 to December 2010, *Time* magazine alone ran over one thousand stories (in print or online) that mentioned oil and gas prices—that the national mood seemed to be tethered to the price of "the Devil's excrement." As one peakist investor in his early sixties reported: "As the price of oil rose, I began to investigate the causes of the price increases." That investigation eventually lead him to peak oil.[47]

In a CBS poll taken just after Hurricane Katrina (September 2005), nearly nine in ten Americans said that their lives had been affected by rising gas prices.[48] In my surveys, one of ten peakists said that these specific increases pointed them to the peak oil theory. A trucker in his fifties who spent $15,000 a month on diesel gasoline, for example, reported that the "!!!!!@#$%&^*!!! price of diesel" after Katrina steered him toward peakism. A Virginia woman in her thirties said that the "gas prices [*sic*] spike after Katrina" led her to "reconsider the status quo," while a liberal in his early thirties, who had "been watching the steady climb in gas prices since 2000, when Katrina spiked prices over $3," found in peak oil the explanation for this increase: "Earlier in the year I had discovered the writings of James Kunstler. I went back to his Web site and others and felt I discovered the missing element in my thinking. . . . The logic of Peak Oil and [M.] King Hubbert seemed sound to me."[49]

Environmentalism and Climate Change

Anxiety and concern about climate change was the third development that led Americans to the peak oil movement. The threat of anthropogenic climate change has been on the public radar since 1988, and as early as early as 1991 35 percent of Americans said that they were "a great deal . . . worried" about global warming.[50] However, until very recently climate change consistently rated at the bottom of the hierarchy of public concerns, far behind the economy, unemployment, foreign policy, and health care. This seemed to be changing by 2005–7, however.

Scientific evidence of a general warming trend, drought, and even extreme weather events has only so much impact on concerns about climate change, and popular culture has been especially significant in shaping the debate over climate change. As such, we might appropriately date this renewed millennial concern to Roland Emmerich's 2004 film *The Day After Tomorrow*, which portrayed abrupt climate change that results in an ice age almost overnight and sparked a newfound interest in the potential immediate (rather than future) impact of climate change in newspapers and blogs.[51] A year later, Hurricane Katrina, which 68 percent of Americans associated with climate change, seemed to present a turning point for climate awareness.[52] Previously, many Americans had little firsthand knowledge of the potential negative consequences of carbon emissions since, as Michael Ziser and Julie Sze pointed out, "the most tangible evidence of warming on U.S. soil was to be found in coastal arctic villages."[53] If Katrina brought the potential consequences of climate change home, former vice president Al Gore's film of the following year, *An Inconvenient Truth* (2006), was just as influential, presenting clear science, stark predictions, and arresting images such as the flooding of an aerial map of Manhattan.[54] As a work of propaganda the film was successful: a Nielsen poll found that 89 percent of American viewers said that they were more aware of climate change's causes and consequences because of it, and 74 percent said that they had begun taking individual actions after viewing it.[55]

The year 2008 seemed like a watershed. The UN Intergovernmen-

tal Panel on Climate Change (IPCC), an international and widely
respected organization composed of thousands of scientists, released
a comprehensive report on climate change that seemed to finally put
to rest any doubts about the existence of a scientific consensus; Al
Gore shared the Nobel Peace Prize with the IPCC; and MTV sponsored
a "Live Earth" concert that spanned all seven continents. Fifty-nine
percent of Americans told pollsters that "it's necessary to take major
steps starting very soon."[56] From Hollywood films with environmental
messages (in 2008 alone: *Wall-E*, *The Happening*, and *The Day the Earth
Stood Still*) to corporate greenwashing campaigns, a sea change in environmental concern seemed to be at hand.

Many peakists were led to anxieties about resource depletion
through their research into climate change. A computer programmer reported, "I learned about peak oil while I was studying climate
change," while a retired Denver meteorologist noted, "[I] got interested
in peak oil ancillary to my environmental studies. I just happened to
catch a couple of authors who had written peak oil books speaking
to the environmental issue and picked up on the concept." An Idaho
man said that "abnormally warm weather during the winter upset" his
winter ski plans, which "led [him] to getting informed about climate
change," which "led to [his] getting informed about our dependence
on fossil fuels," which "led [him] to peak oil."[57]

Abundance, Scarcity, and Limits in Recent American History

While the Iraq War, rising gas prices, climate change awareness, and
Internet access provide the proximate causes of the peak oil phenomenon, these events and developments unfolded amid shifting popular
conceptions about energy availability, the free market, and technology.
Recent concerns about oil depletion must also be located in the series
of historical cycles of beliefs about abundance, scarcity, and limits in
recent American history.

Confidence in abundance and a general optimism are certainly consistent with long-standing mythology about the "national character"
of Americans. From early chronicles of the United States, such as Alexis
de Tocqueville's classic *Democracy in America* (1835) through the 1970s,
Americans have traditionally been cast as optimists whose vision of
the future was as boundless as their expanding western border. While
conducting researching for his novel *Amerika* (1927), and knowing the
United States only through its literature and reputation, Franz Kafka

told his compatriots: "I like the Americans because they are healthy and optimistic."[58] Half a century later, in a speech given in Washington, DC, on January 20, 1961, John F. Kennedy asserted that "the American, by nature, is optimistic," and few would have disagreed.[59]

However, the role of "nature" in this formulation is more interesting than Kennedy may have realized. To what extent was twentieth-century American optimism the product of or reliant on natural abundance, both real and imagined? Throughout the last century, an underappreciated determinant of what one might call the *national mood* was an estimation of the abundance or scarcity of natural resources, especially petroleum. Cheap oil played a crucial but often unacknowledged role in the development of American geopolitical power and its citizens' high quality of life in the post–World War II period, essentially mixing an expectation of unlimited cheap energy into the foundations on which modern American life was constructed. In the late 1960s and the 1970s, this expectation was challenged by environmentalists, whose concerns about limits to continued growth were seemingly corroborated by environmental disasters, oil embargoes, population growth, and economic woes. In the 1980s, behind Ronald Reagan's optimism and a global petroleum glut that led to low prices, these concerns vanished for almost two decades. In the second half of this chapter, I place the peak oil movement in this historical context.

Before tracing the history of growth, abundance, and scarcity in American popular political economic thought, I need to define my terminology. While *growth* usually denotes a significant increase in economic production and consumption, the term has assumed connotations far beyond the economic sphere. It has come to signify human progress itself, with attendant industrial development, material consumption, technological innovation, and expansion of individual freedom. Economic growth is a pillar of the dominant social paradigm identified in the last chapter, but it was not always so. The idea that progress, national strength, and even general happiness are the necessary by-products of an indefinitely expanding economy, measured as the percentage rate of increase of GDP from one year to the next, is a nearly universal assumption of contemporary public discourse with which few Americans seem to disagree. But it would have struck most people throughout human history, including the first three decades of the twentieth century, as a strange notion indeed.[60]

The Great Depression is the most appropriate place to begin any discussion of modern notions of growth, which one historian has correctly identified as "easily the most important idea of the twentieth century."[61] During the 1930s most politicians and economists promoted economic "balance" and "security," not unfettered economic growth, while the federal government adopted policies intended to avoid economic "stagnation," such as cutting farm production to steady consumer prices. Three decades after the historian Frederick Jackson Turner declared that the western frontier was closed, many economists warned of the nation's maturity and impending decline, while some southerners feared that continued industrial development and modernization would weaken their communities and deliver power and wealth to distant interests.[62] After Pearl Harbor, however, ambivalence about economic growth disappeared as the entire industrial machine was put toward the war effort. Historians have called World War II a "gross national product war" with good reason, as it ultimately turned not so much on strategy or moral rectitude as on which side could produce more matériel (tanks, trucks, and aircraft) and devastating new weapons.[63]

During the war "American society seemed to have discovered the secret of economic progress," as Godfrey Hodgson put it, and in its aftermath few were interested in returning to the modest goals of balance and stability.[64] In the immediate post–World War II period, then, economic advisers and government economists formulated a lasting political philosophy that Robert M. Collins has termed *growthmanship*, raising growth "from an overriding economic goal (first among equals) to a new organizing principle for a neo-corporatist political economy."[65] By his 1949 State of the Union address, President Truman could announce that it would not be enough to "float along ceaselessly" or "merely to prepare to weather a recession if it comes." Instead, the country must "constantly . . . achieve more and more jobs and more and more production," which would "mean more and more prosperity for all the people."[66]

In the 1950s, conditions seemed to bend their will to this new ideology. The baby boom (1946–64) constituted the largest population increase in American history, and GNP per capita increased 24 percent between 1947 and 1960, while personal consumption spending increased by 22 percent. Behind strong labor unions, the middle class swelled. With only 7 percent of the global population, the United States now accounted for half of global manufacturing output and accrued nearly half of the world's income each year.

In a historical moment in which western mythology played a considerable role in the national imaginary, economic growth was cast as a new and limitless frontier.[67] Alvin Hansen, a leading Keynesian economist, declared consumption to be "the frontier of the future."[68] Notions of progress, linked to American exceptionalism and Manifest Destiny, evolved (or were slotted) into the theory of economic growth, which relied on and was expressed through material consumption. In this vein, Lizabeth Cohen has described the "new postwar ideal of the purchaser citizen who simultaneously fulfilled personal desire and civic obligation by consuming."[69] This consumption ethic is evident in some of the era's leading nonfiction. In his 1954 *People of Plenty*, for example, David Potter wrote that society expected the modern man to "consume his quota of goods," "of automobiles, of whiskey, of television sets," and "regards him as a 'good guy' for absorbing his share, while it snickers at the prudent, self-denying, abstemious thrift that an earlier generation would have respected."[70] In *The Organization Man* two years later, William Whyte noted that "thrift is becoming a little un-American."[71]

If Dwight Eisenhower seemed reluctant to promote or rely on growthmanship, John F. Kennedy was not.[72] The pursuit and expectation of economic growth became one of the key features of his New Frontier, an official imperative epitomized by his creation of the cabinet Committee on Growth. The signs were everywhere, sometimes quite literally—soon after his inauguration, placards asking "What have you done for Growth today?" were placed at every desk of the Department of Commerce.[73]

While growth seemed, in this formulation, an end in itself, it actually served two key political functions during this period. First, economic growth became a key site of competition in the Cold War. As the Soviet Union and the United States sought to attract nonaligned states, their ability to produce cheap goods and, thereby, deliver material comforts and a high quality of life became a crucial issue. This consumption race was embodied in the famous "kitchen debate" between Nikita Khrushchev and Richard Nixon in 1959.

Second, economic growth played (and continues to play) a vital role in allowing American politicians to sidestep questions about the distribution of wealth, which had become a divisive issue in the 1930s. In a metaphor of the era, the question of how to slice the pie of national wealth (between different classes and racial and ethnic groups) could remain unanswered so long as the pie continued to grow. National stakeholders that might have offered a differing vision followed this lead: unions, for example, shifted their focus from solidarity, power, and structural changes to collective bargaining over wages and benefits. As Hodgson observed, it became "tempting to jump to the conclusion that there was not much more to democracy," or, for that matter, politics, "than keeping unemployment under 5 per cent, guaranteeing a swelling flood of consumer goods to the employed, and increasing the gross national product by a respectable percentage each year."[74] If postwar technocratic politics presented itself as a science, with similar universality and authority, we might view growthmanship as a new form of alchemy through which finite physical resources are transmuted into perpetual expansion.[75]

Expectations of a bottomless pie led to a sense of expansive national possibilities. As the historian Lawrence R. Samuel observed, post–World War II American "futurism was, in a word, limitless," envisioning a "consumer utopia" whose "sheer abundance [was] capable of solving all of our social ills."[76] Amid an unprecedented boom in population growth, economic output, health, material consumption, and quality of life, it is no surprise that, as James T. Patterson has noted, "the majority of the American people during the twenty-five or so years following the end of World War II developed ever-greater expectations about the capacity of the United States to create a better world abroad and a happier society at home."[77] The postwar ideology of growth, seemingly validated by an ever-rising GDP, enabled the country's expansive Cold War foreign policy, including deep-pocketed support for allies and regular military interventions around the globe. At home, the sense of unlimited national potential facilitated Eisenhower's pragmatic acceptance of New Deal programs and buttressed Lyndon Johnson's pursuit of the Great Society. In keeping with the gospel of growth, these programs would not redistribute wealth but simply democratize access to the American way of life and patterns of consumption.

Beyond politics and foreign policy, a slew of technological innovations changed the way that Americans lived and the futures they imagined. Research and development for World War II and Cold War technologies produced regular innovations that gave many Americans a sense of invincibility.[78] The atom had been harnessed and might provide cheap, safe energy for thousands of years. Medical research produced antibiotics, vaccines, pacemakers, and, in 1960, the birth control pill. After the shock of Sputnik's orbit, the United States soared ahead in the Space Race—the moon was the next stop and then perhaps Mars, if not even more distant destinations. Drawing on the popular appeal of technological marvels that opened new, unconsidered possibilities, science fiction moved into the mainstream. Presenting utopian visions of a grand tomorrow, futurists like Carl Sagan and Alvin Toffler, the space art of Arthur Radebaugh, and television shows like *The Jetsons* (1962–63) represented a projection of growthmanship into the future.[79]

As Brian C. Black noted, "the rise of technological progress" in post-war America "had convinced most people that human ingenuity could overcome all problems that came from limited resources."[80]

While human ingenuity played a vital role in America's post–World War II economic and geopolitical growth, so did petroleum. As Eisenhower admitted in his 1955 State of the Union address: "At the foundation of [America's] economic growth are the raw materials and energy produced from [its] minerals and fuels."[81] The key ingredient in the prosperity and optimism of the postwar decades was cheap oil—from 1945 to 1970 the per barrel price of crude oil remained below $20.00 (in 2009 dollars; see fig. 8). It enabled the rise of the automobile, which became the signature commodity of the postwar years as the number of cars owned by Americans doubled between 1945 and 1950. This vast fleet traveled on roads made mostly of asphalt, a petroleum derivative, courtesy of the 1956 Interstate Highway Act, which constructed forty-one thousand miles of freeway. Cars and roads produced elemental transformations that fundamentally altered the structure of everyday life, such as decentralized suburbs and exurbs, long commutes to work, and a national drive-in culture. Assuming an endless supply of cheap

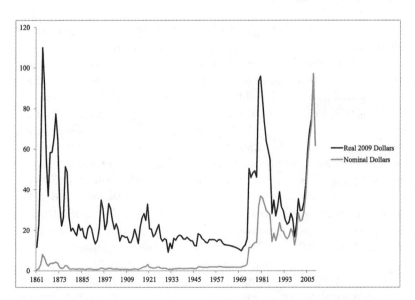

FIGURE 8   Crude oil prices, 1861–2010. *Source: BP Statistical Review of World Energy* (June 2010), http://engineering.dartmouth.edu/~d30345d/courses/engs41/BP-Statistical Review-2010.pdf.

fuel, suburban developers and advertisers promoted suburbia as the ideal. By 1960 more Americans lived in suburbs than in cities.

While many Americans have long considered the post–World War II period to be an exceptional golden age, environmental historians have recently highlighted just how exceptional it was.[82] We should recall that times of scarcity have been an expectation and concern for almost all human history and prehistory. As Collins put it, "in all prior civilizations and social orders, the vast bulk of humanity had been preoccupied with responding to basic material needs," but, in the mid-twentieth century, "the age-old bonds of scarcity were broken," marking a "fundamental change in the human condition."[83] This leap became possible because of technological innovations, to be sure, but also because petroleum is a unique substance, formed over millennia, the likes of which human beings had never discovered and most likely never will discover again on this planet. Indeed, the ubiquity and flexibility of petroleum are somehow mysterious, almost magical. Imre Szeman observed that, "despite all that has been and continues to be written about oil, it still seems to be difficult to capture the fundamental way in which access to petrocarbons structures contemporary social life on a global scale," and Michael Ziser noted, in a similar vein, that "our Western bodies, homes, communities, governments, arts, and ideologies are—in a sense that is still mysterious and infrequently spoken of—expressions of the mystical surplus of energy that is fossil fuel."[84] Floating in an ecology of oil, it is easy to forget just how unique it is.

As a historical event, the exploitation, refinement, and application of petroleum on industrial scales in the post–World War II period are perhaps best understood side-by-side with similar historical events. The closest parallel is to be found not in whale oil, coal, or hydrofracking but in the arrival of human beings on new land masses, such as North America (approximately 15,500 years ago) and Australia (approximately 40,000 years ago). Archaeologists have found that, in these cases, prehistoric pioneers also discovered a seemingly abundant source of energy in the highly caloric flesh of large mammals lacking natural defenses against Homo sapiens. In almost every case, they exploited this new source of energy quite inefficiently, driving entire herds over cliffs, for example, as well as carelessly, hunting them to extinction within a century.[85]

Petroleum is just as unique as the giant ground sloths that once traversed North America because of two of its properties: high energy density and transportability. As all sources of energy (except geothermal) are essentially concentrations of solar radiation converted and stored through photosynthesis, their energy density reflects how much sunlight they contain. For millennia, humans consumed only sunlight that had recently fallen on plant matter or, when those plants were consumed by fauna, accumulated in animal fat. Preindustrial societies had two primary sources of energy: human and animal labor and biomass, especially wood (when burned, mature trees release the energy equivalent of decades of sunlight). Burning coal, which is sunlight transferred through photosynthesis into plants, then compressed in high temperature and pressure over millions of years, constituted a historic step forward for human beings since for the first time we were drawing on an ancient bank of energy. But even algae and zooplankton contain far more concentrated sunlight than plant matter does, and petroleum is composed of millions of years of such organisms, a seemingly limitless bank account of ancient sunlight compressed by geologic forces into a dazzlingly high energy density (see fig. 9). Almost as important is petroleum's liquidity, which makes it easily transportable and able to flow through inexpensive pipelines as coal never could.

As Clive Ponting has pointed out, until very recently "all societies faced an energy shortage" that effectively limited their ability to grow, in terms of population, or what we now call *economy*, as well as the quality of life available to any more than a small number of (elite)

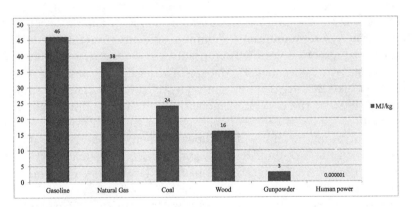

FIGURE 9   Petroleum is more energy dense than human power, wood (and other biomass), and coal, and the ability to harness an ancient bank of sunlight through it led to an improved quality of life for hundreds of millions of human beings in the twentieth century. As Bradley E. Layton put it: "Gasoline is ten quadrillion times more energy-dense than solar radiation, one billion times more energy-dense than wind and water power, and ten million times more energy-dense than [ideal] human power." Bradley E. Layton, "A Comparison of Energy Densities of Prevalent Energy Sources in Units of Joules Per Cubic Meter," *International Journal of Green Energy* 5, no. 6 (2008): 438–55, 441.

people.[86] By drawing on an ancient bank of concentrated sunlight through petroleum and, thus, adding an unprecedented augmentation to their energy consumption, Americans and other Westerners in the mid-twentieth-century entered a historically unique situation that J. R. McNeill has described as "a liberation from the drudgery of endless muscular toil and the opening up of new possibilities well beyond the range of muscles."[87]

By the twenty-first century, Americans lacked firsthand experience of the transition from this mostly "somatic" energy regime to the current regime of fossil fuels. The historical uniqueness of the post–World War II world—and, indeed, of the entire twentieth century, in which more energy was deployed than in all previous human history combined—has long been forgotten, its petrofoundation taken for granted.[88] The way of life enabled by this unprecedented bonanza became the new normal, mass consumption no longer conspicuous but expected. From a broader perspective on human history, however, we should recall that it is abundance that is exceptional and that scarcity is hardly an empty threat.

Limits-to-Growth Environmentalism and the 1970s Oil Shocks

From this deep historical perspective, the concerns about energy that emerged in the late 1960s and the 1970s were hardly fanciful. Though they may now seem unusual, the kinds of anxieties that motivated the emotional responses we saw from peakists in the last chapter were at the center of environmental concern in the United States only four decades ago. Historians have rightly identified Rachel Carson's *Silent Spring* (1962) as the catalyst of modern American environmentalism, but scholars such as Thomas Robertson have amended this narrative by highlighting the unprecedented public attention that the so-called Malthusian tracts of the late 1960s and early 1970s drew to environmental issues.[89] These concerns are named after the British economist Thomas Robert Malthus, whose *Essay on the Principle of Population* (1798) argued that future population growth would inexorably outstrip food supply, causing poverty, epidemics, wars, and general misery. Although his work was read widely in nineteenth-century Europe—Charles Darwin, for example, considered it an influence on his theory of natural selection—it had much less of an impact in the United States, where Thomas Jefferson's vision of an expanding republic with limitless land and opportunities held sway.

In the late 1960s, faith in and expectation of perpetual economic growth underwent a sudden reversal. By this time, population growth was already making headlines through its sheer numbers—global population, two billion in 1927, hit three billion in 1960 and was expected to surpass four billion in the early 1970s, marking the first time in human history that a generation had seen the global population double (and been aware of it). In this environment, Paul Ehrlich's 1968 bestseller *The Population Bomb* ushered in a Malthusian moment in American culture that seems inconceivable today. Updating Malthus's pessimistic prophecies in a (melo)dramatic style, Ehrlich argued that the planet's environmental crisis boiled down to one fundamental issue: too many human beings competing for a limited set of resources. The impact of this idea was bolstered by a cacophony of similar claims from other authors, such as the ecologist Garrett Hardin, whose influential essay "Tragedy of the Commons" (1968) suggested that the United States would not be spared the consequences of the degradation of global environmental commons.[90] Historic international events added their own momentum to these apprehensions, such as Apollo 8's space mission of the same year, which broadcast the first images of the entire planet Earth from outer space, encouraging a new conception of the planet as one interconnected, finite, and potentially fragile ecosystem.

Malthusian environmental concerns moved into the public arena and were called into service by diverse political actors for disparate purposes. As its title suggests, *The Population Bomb* dramatized the threat of overpopulation by framing the issue in classic Cold War terms of weapons and interdependence. Overpopulation in the third world would lead to scarcity and poverty, which might provide fertile ground for the growth of Soviet communism in nonaligned nations. In this way, the "population bomb" was just as immediate a threat to American interests as nuclear weapons were. At the same time, feminist activists repurposed Malthusian anxieties for their own ends, citing overpopulation as another reason that women should have legal access to contraception.[91] Most importantly, *The Population Bomb* raised the stakes of environmental awareness. Although environmental issues already had broad public support, Malthusian concerns "set the apocalyptic horizon of environmentalism," as John S. Dryzek noted, "giving the basic reason why care and concern about the environment were not just desirable, but also necessary."[92] *The Population Bomb* was a major influence on the architect of the first Earth Day, Wisconsin senator Gaylord Nelson, who even placed an article by Ehrlich into the *Congressional Record* in 1970.[93] Whereas *Silent Spring* focused primarily on the influence of specific chemical contaminants on wildlife, neo-Malthusians argued that the very future of the human species was at stake. The abrupt shift away from post–World War II growthmanship was remarkable—in a 1979 survey, 31 percent of Americans described themselves as "anti-growth," with another 39 percent "highly uncertain."[94]

The second major work of Malthusian environmentalism was 1972's *Limits to Growth*, which Max Lerner correctly predicted would "detonate through the debates of the '70s."[95] *Limits to Growth* argued that finite natural resources—from petroleum to heavy metals to arable land—would impose an eventual limit on the growth of contemporary human civilizations and predicted a dramatic "overshoot and collapse" if the course were not altered. Written collectively by an organization composed primarily of scientists called the Club of Rome, part of the book's appeal was its novel use of computer modeling, expressed in charts and graphs that peak oil authors would emulate three decades later. *Limits to Growth* sold four million copies in its first four years, and the report and its reception made headlines in national newspapers for months.[96]

For a short time, *Limits to Growth* pushed Malthusianism back into the center of public debate. Many reviewers disagreed with specific contentions, but even conservatives granted its basic claim that economic systems expecting infinite growth (namely, capitalism) were incompatible with natural resources that are finite—*Business Week*, for example, commented that, "for all the criticism, practically everyone agrees that on a finite planet, growth must end sooner or later."[97] Newspapers published articles such as a 1973 *Chicago Tribune* cover story with the headline, "Fuel, Resources Dwindling: Can America Survive the 20th Century?" In it, the Pulitzer Prize–winning journalist Ronald Kotulak asserted that "our lives, our economy, and our country revolve around the grossly wasteful use of abundant energy" but that, since "fossil fuels, except for coal, are running out," the *Limits to Growth* prediction—described here as a future "marked by an industrial flame-out, breakdown of society, and a soaring death rate" until a global "collapse around the year 2100"—must be heeded.[98] Questioning the value of economic growth became widespread. Nixon's top environmental adviser, for example, made headlines in 1972 by calling for a "national debate on the desirability of growth," noting that "the finite nature of man's physical resources calls for systematic constraints on expansion to avert the degradation of 'the quality of life.'"[99] The impact of this Malthusian moment on average Americans could be measured even three decades later by the number of older peakists who cited both *The*

Population Bomb and *Limits to Growth* as books that significantly influenced their lives. A New Jersey man, for example, explained his interest in peak oil by reporting: "Way back in 1972 I read *Limits to Growth*, which was pivotal in my environmental leanings."[100]

In this atmosphere, the 1973 oil embargo seemed to validate the limits-to-growth thesis. In response to Israel's attack on Syria in October 1973, OPEC unsheathed its "oil weapon" by ceasing exportation of petroleum to countries that supported Israel, including the United States.[101] The effect was immediate: the 1973 oil embargo led to a fourfold increase in the price of gasoline and major shortages of heating oil. Long lines at gas stations, hoarding, and impatient, panicked, and even violent motorists compelled President Nixon to impose a number of national conservation measures that seemed to presage, if not register, a national decline, such as lowering highway speed limits, cutting air travel, and even darkening national monuments in Washington, DC.[102] Occurring in the same year as the Paris Peace Accords and the humbling return of US forces from Vietnam, the embargo seemed to be a sign that Americans no longer enjoyed the power and freedom to do as they pleases, in the international political arena as well as on their local highways. Thirty years later, some peakists still considered the embargo to be one of the most influential events in their lifetime. Of all survey respondents who were at least five years old at the time of the 1973 embargo, one out of four cited it as the event that triggered their eventual conversion to peakism. Indeed, this experience of scarcity and limits, so unfamiliar to Americans born since approximately 1980, is one potential explanation for the advanced average age of peakists (forty-seven) in comparison to that of participants in other social movements.

Scholars also weighed in on the contributions of culture and politics to encouraging unsustainable growth, and economists such as Herman Daly began outlining what an economic system based on a steady state might look like. E. F. Schumacher's 1973 nonfiction *Small Is Beautiful*, which promoted a "maximum of well-being with a minimum of consumption," became a best seller even before President Jimmy Carter invited its author to the White House.[103] In environmental scholarship, more attention was paid to the potential impact of resource scarcity on society than to the impact of human societies on the natural environment. As late as 1976, one state of the field described "the dominant theme" as humanity "entering an era of ecological limits."[104] By the late 1970s, of course, the sense that that decade was an age of limits went all the way to the top—shortly after leaving office in 1981, Carter opined that "dealing with limits" had been the "subliminal theme" running through his presidency.[105]

As quickly as these Malthusian concerns appeared, so too did they vanish from public discourse. By the mid-1980s, "proponents of environmental apocalypse" seemed, as Frederick Buell observed, like "a bunch of monotonal, gloomy, anti-human, neo-Malthusian purveyors of doom."[106] Part of this caricature was sketched by conservative antienvironmentalists, but there were concrete reasons that limits to growth wore out its welcome. First, an acknowledgment of the limits to continued growth demanded radical, expensive changes to American life that would need to be centrally planned. While some younger Americans were open to (and, indeed, embraced) these kinds of changes, other groups, such as corporations, businesspeople, and most politicians, were not quite as open to transforming capitalism. In comparison, other environmental paradigms, such as "ecological modernism" and "sustainable development," maintained that environmental protection might be perfectly consistent with sustained economic growth and, thus, proved more appealing to vested interests.

Second, a number of environmentalists who believed that humanity was nearing its limits to growth voiced support for measures that verged on authoritarianism and even racism. Paul Ehrlich and others sometimes focused their concerns about overpopulation on third world nations with high birth rates, especially India, and *The Population Bomb* seemed to blame some of the urban uprisings of the late 1960s on overpopulation. At times, these supporters even suggested that forced sterilization might become necessary. Third, while authors such as Ehrlich and Donella Meadows (of *Limits to Growth*) intended their forecasts to be catalysts for social reform and policy modifications, their critics in the mid- and late 1970s focused on whether individual, short-term predictions bore fruit. When they did not, this approach lost credibility.[107] As Thomas Robertson put it, the "supercharged" rhetoric of limits to growth "sold books and spurred people to action, but also opened environmentalists to claims of being 'chicken littles' and misanthropic, even racist, authoritarians."[108]

The election of Ronald Reagan in 1980 was a death knell for the mainstream discourse of limits. In the 1960s and the early 1970s, Reagan, then governor of one of the country's most environmentally sensitive states, preached concern for the environment, even organizing the Governor's Conference on California's Changing Environment in 1969. But by 1980 he and perhaps the general public had changed their tune. After a decade of being reminded of limits—to growth, to personal consumption, to military influence, and to national power—many Americans embraced Reagan's optimistic proclamation of a new "morning in America." Claiming that predictions of doom were often wrong, he mocked the Carter administration's wariness during the 1980 campaign and promised in his inaugural speech that "we're not, as some would have us believe, doomed to an inevitable decline."[109] On taking office, he ordered the immediate removal of the photovoltaic cells that Carter had installed on the White House roof and proceeded to strip all federal environmental initiatives of funding, power, or both.[110]

Faced with an administration and a general public with less interest in green issues, environmentalists replaced the limits-to-growth perspective with a less dramatic paradigm of change, known as *global environmental change*. In the wake of erroneous Malthusian predictions, environmental groups and environmentally minded academics and authors were now more dependent on scientific justifications for actions, and in the 1980s scientists point to the destructive human impact on the natural environment, such as water contamination, acid rain, ozone depletion, and deforestation.[111] By the mid-1980s, then, environmentalists were again highlighting what "humans were doing to the environment, as was true when environmental quality emerged as a social problem in the late sixties," although the apocalyptic weight of the limits-to-growth paradigm had added a global imperative to these issues.[112]

In this paradigm of global environmental change, energy issues surfaced mostly in the case of toxic spills, such as the *Exxon Valdez* spill in 1989. With the brief exception of the Gulf War, energy issues faded from the public eye in the 1980s and 1990s. Once again, the price of oil was a bellwether: between 1979 and 2000, the average national price of a gallon of gas hovered between $0.94 and $1.40 (adjusted for inflation), with a small jump in 1990–91 during the Gulf War.[113] The high and unpredictable prices of the 1970s created an incentive to search for new fields and develop new drilling and refining technologies, leading to the diversification of imported oil enabled by new sources, especially from Mexico, the North Sea (England, Norway, Denmark, and Germany), and, in the 1990s, Russia.[114]

By the 1990s, abundance and optimism were once again the order of the day. Gas prices were low and stable, the United States firmly in control of global political hegemony and far removed from fears of scarcity. Behind the neoliberal stewardship of the Clinton administration and the financialization of the US economy, the expectation of continued economic growth now assumed the form of collective investment in the stock market. Few questioned the alchemy of growthmanship, so relatively few wondered whether the meteoric rise of Internet and technology stocks and home prices could continue indefinitely. Larry Summers, for example, then serving as Clinton's secretary of the Treasury before becoming Barack Obama's chief economic adviser, declared emphatically that the Clinton administration would not "accept any 'speed limit' on American economic growth."[115] The motivation industry described in the last chapter, buoyed by the new school of positive psychology, became a cultural force, with many corporations training their workers to be positive team players. Techno-optimists argued that computer technologies had created an American "productivity miracle" and predicted this trend would continue deep into the "roaring" first decade of the twenty-first century.

This history of the ebb and flow of beliefs about abundance and scarcity helps us understand the reactions that many peakists provoked in the last chapter. Concerns over limits to economic growth, capitalism, and American power seemed to have disappeared from the national discourse in the early 1980s but reappeared circa 2006 around peak oil and climate change. As a peakist in her late fifties said: "I started trying in '74 to talk to people about resource depletion, and people thought I was out of my mind, so I quit discussing [it] until 2005, when it became acceptable in some circles."[116]

Crying Wolf: The Legacy of Malthusian Environmentalism

In 2002, when Exxon Mobil sarcastically commemorated the thirtieth anniversary of *Limits to Growth* in a print advertisement, its assertion that the Club of Rome was "wrong" and rejection of the idea that "our current lifestyles and use of raw materials are unsustainable" would have been accepted by most Americans.[117] By the 1990s, these and other Malthusian environmental concerns were considered distant historical events worthy of a smirk, if not leftist attempts to impose some kind of environmental autocracy. If my students are at all typical, most Americans born since the early 1980s are not even aware of the Malthusian moment of the late 1960s and early 1970s, so thoroughly was the doctrine of economic growth reasserted and accepted in the 1980s and 1990s by liberals and conservatives alike.

Nonetheless, this historical moment and its seemingly failed predictions cast a long shadow and constitute one framework for understanding the failure of scientists and environmentalists to convince Americans to respond to warnings about climate change as well as resource

depletion. The stereotype of the environmental apocalyptic prophesying the end of the world (fig. 10) has become a cultural archetype and
implicitly breeds skepticism about the most recent warnings. So it is
that one legacy of America's Malthusian moment was widespread environmental doubt, which conservatives and antienvironmentalists have
exploited at every opportunity. In a 2001 *Wall Street Journal* op-ed, for
example, Philip Stott could argue that global warming was merely a
"new myth [that] was seen to encapsulate a whole range of other myths
and attitudes that had developed in the 1960s and 1970s, including
'limits to growth,' sustainability, neo-Malthusian fears of a population
time bomb, pollution . . . and an Al Gore–like analysis of human greed
disturbing the ecological harmony and balance of the earth."[118]

A discourse of limits, closely tied to the contemporary fossil fuel
energy regime, returned to the mainstream in the first decade of the
twenty-first century. Although most advocates of peak oil tended
to stress oil depletion above all other issues, a surprising number of
mainstream environmentalists subscribed to peakism. For example, in
Eaarth: Making a Life on a Tough New Planet (2010), Bill McKibben quotes
liberally from the peakist author Richard Heinberg, describes the long
decline of energy returned on energy invested (EROI) for global oil production, cites James Schlesinger as saying that "the battle is over, the
oil peakists have won," and suggests, along with most peakists, that
one cause of declining home prices in suburbia is the rising cost of gasoline.[119] In a discussion of how the petroleum industry is responsible
for one-quarter of leaked toxic materials in the United States, the 2010
second edition of Sandra Steingraber's classic *Living Downstream* (1997)
asserts that global oil production "has peaked, or is peaking, or will
peak soon."[120] In 2007, James Hansen, one of the world's most influential and well-known climatologists, coauthored an article on the potential impact of peak oil scenarios on future climate change.[121]

More importantly, mainstream environmentalists began returning
to a discourse of limits, especially in describing climate change. Desperately seeking a way to communicate the gravity of the climate crisis,
which already causes an estimated 300,000 deaths every year, environmentalists turned to limits and depletion as one strategy.[122] In this
conceptualization, the limit to growth—or even survival—is not an energy source but an energy sink. Carbon sinks, such as oceans, forests,
and soil, sequester carbon dioxide emitted from fossil fuel consumption and, thus, provide a crucial buffer against even more rapid climate
change. Scientists and environmentalists began warning that the ability of carbon sinks around the world to store carbon dioxide, previously

FIGURE 10   A cartoon in the *Washington Times* mocks Al Gore's September 2011 twenty-four-hour Web cast "24 Hours of Climate Reality," which highlighted the global consequences of anthropogenic climate change. Alexander Hunter, *Washington Times*, September 16, 2011, http://www.washingtontimes.com/news/2011/sep/16/gores-bore-a-thon/#ixzz2TlVdELcK. The accompanying article concludes: "The more Mr. Gore speaks, the more the public realizes the only thing warming up is the hot air surrounding the climate change issue." The cartoon identifies him in the tradition of the sidewalk apocalyptic whose predictions have always failed to materialize and, thus, always will. The inclusion of a Ron Paul figure shows that environmental claims are merely another form of bogus apocalypticism.

not widely understood, might be declining, leading to a faster accumulation of atmospheric carbon dioxide than had been expected.[123]

By 2012, the gravity of climate change motivated mainstream environmentalists to return full circle to the mathematical predictions of *Limits to Growth*. Bill McKibben kickstarted his organization 350.org's "Do the Math" campaign with a widely read article in *Rolling Stone*. Since "pictures of those towering wildfires in Colorado" and "the size of your AC bill" had not convinced the average reader of the urgency of action, McKibben delivered "three simple numbers that add up to global catastrophe," along with the kind of specific predictions that environmentalists had eschewed since the 1970s. He computed that, if oil companies were to extract and sell their 2,795 gigatons of proven oil and gas reserves, worth $27 trillion, the resulting warming would exceed what scientists consider the threshold of tolerable warming, 2 degrees Celsius (3.6 degrees Fahrenheit), by a factor of five. The purpose of engaging this limits discourse was to erase the frustrating requirement of emphasizing likelihood and probability—risk, above all—instead of certainty in predicting the effects of our actions from environmental communication. "Before we knew these numbers," McKibben wrote, "our fate had been likely. Now, barring some massive intervention, it seems certain."[124]

McKibben's clarion call was cogent, persuasive, and moving, relying on scientific consensus to deliver an emotional appeal for action, just as *The Population Bomb* and *Limits to Growth* had in their time. However, those earlier jeremiads—or, rather, the widespread misconception of their messages—had damaged the credibility of environmentalists. McKibben, Gore, and other scientists and environmentalists were simply slotted into the long line of sidewalk prophets with failed predictions.

Limits, Regulations, and Markets

If historical connotations have tarnished the discourse of limits in the United States, we should remember that limits themselves are fundamental to any environmental perspective. As Richard McNeill Douglas put it: "While environmentalism means different things to different people, almost all its forms are united by a central core premise: that nature imposes physical limits to human consumption."[125] In the 1980s and 1990s, environmentalists continued to speak of *limits*, but they shifted their rhetoric to one of *regulation* and *protection*. As our historical perspective on abundance and scarcity in twentieth-century American history shows, an opposition to limits is partially psychological. An opposition to regulation or environmental protection, as environmentalism was often framed in the 1980s, is another matter.

The connections between abundance and scarcity, on the one hand, and regulation and free market neoliberalism, on the other, are not hard to tease out. In response to the new wave of regulation in the 1960s and the early 1970s, such as the creation of the Occupational Safety and Health Administration and the Environmental Protection Agency, large corporations began to systematically organize and lobby against regulatory expansion. They did so primarily by borrowing arguments advanced by emerging libertarian think tanks about the magic of the free market, which would, in their perfect state of affairs, require no regulation whatsoever. But, while true market fundamentalists might argue that the free market can manage scarce resources better than any government regulator, this claim did not carry the same weight for the average American. Crucial resources that are also scarce seem to demand government intervention of some sort; resources that are limitless, however, require none.

The tension between environmental awareness, which by necessity recognizes and respects the limits of the planet, and a belief in "America unlimited" is not a historical or a hypothetical question but a factor in our lack of response to contemporary environmental challenges, including climate change. If this balance seemed to be moving one way during the Malthusian moment of the early 1970s, by the 1980s the scale had been pushed in the opposite direction by the heavy finger of a previously fringe, now-ascendant political philosophy of libertarianism. With wealthy backers and a champion in Ronald Reagan, corporate interests, libertarians, and neoliberals promoted a resurgent vision of an America without limits and, thus, without any need for regulation.

As we will see in the next chapter, by the time the peak oil movement formed around a vision of impending scarcity, libertarian influence, filtered through decades of talking points about the inefficiency of big government as well as the cyberlibertarianism of the Internet, would have an influence far beyond political philosophers, think tanks, and policymakers.

THREE

Alone Together: The Libertarian Shift and the Network Effect

What depresses me about peak oil is . . . the individualistic survivalist streak in it. AUSTRALIAN MALE, THIRTY-SIX TO FORTY

I tried talking with my husband. He got angry and accused me of being negative and said we're too old to change and we might as well curl up and die. . . . I just keep looking for community—mostly online. WHITE FEMALE, POLITICALLY "PROGRESSIVE," MINNESOTA, FIFTY-SIX TO SIXTY

This is a time for individuals and their networks, not for groups.
BARRY WELLMAN

In this chapter I provide two interconnected explanations for the individualism of peak oil believers described in chapter 1. I view their responses to the threat of environmental crises in the context of a major transformation in American political culture over the last three decades that has not received sufficient attention: the spread of libertarian ideals. Increasingly, Americans of all parties and persuasions conceive of themselves and their citizenship in distinctly libertarian terms (individualistic and private), and this has had a major impact on politics and

This chapter's epigraphs are taken from respondents 9576274, 9601959; and Barry Wellman, "Little Boxes, Glocalization, and Networked Individualism," in *Digital Cities II: Computational and Sociological Approaches*, ed. Mokato Tanabe, Peter van den Besselaar, and Toru Ishida (Berlin: Springer, 2002), 10–25, 11.

social life in the United States. By examining the history of libertarian thought alongside the growth of the Internet over the last thirty years, I argue that this shift has been accelerated by the use of Internet technologies and that the peak oil phenomenon, as a primarily virtual movement, demonstrates the tendency of the Internet to foster a culture of what the sociologist Barry Wellman has termed *networked individualism*.[1] As a result of these two currents, the peak oil movement developed at the start of the third millennium not as a collective social or political movement but as a group of like-minded but isolated individuals. Though peakists formed an impressive virtual community of believers, most seemed to be, in the words of the Internet scholar Sherry Turkle, "alone together."[2]

Collective and Individual Responses

In June 2011, I drove to Sandpoint, Idaho, to visit the Sandpoint Transition Initiative (STI). The second Transition Town in the United States, Sandpoint had been held up as a model of the Transition Initiative, a collective response to peak oil and climate change, by a number of survey respondents. It had even received national press, portrayed in a *New York Times* feature in 2008 as a promising, bustling coalition of energy-inspired communards. The article's author, Jon Mooallem, described the opening event of STI, its "Great Unleashing," and a color photograph depicted thirty dedicated activists meeting on the stage of the baroque Panida Theater. A "new chapter of a worldwide environmental movement" was apparently at hand.[3]

On the Web site of Transition US, the coordinating body for the American Transition movement in the United States, one views a crowded map of official Transition Towns. As of this writing, there were 139 Transition Towns and another 294 groups that were officially "mulling" the proposition.[4] However, the site never clarified exactly what it meant to be designated a Transition Town, beyond having a few committed activists—the phrase *Transition Town* suggests the physical construction of separatist communities or the transformation of urban areas, if not both. Indeed, the Transition Initiative's original endeavor, in Totnes, England, seemed to be well on its way to creating parallel structures that could challenge, supplement, or supersede petroleum-dependent institutions, going so far as to print its own currency.[5] While a number of scholars (mostly in the field of geography) have picked up on the Transition movement, they have primarily analyzed its British incarnations, and few have paid any attention to its low rate of participation in the context of the broader peak oil movement.[6]

I found that the STI was not quite at Totnes's stage of development. According to one member, there were only six people still involved with the central STI body just three years after its founding. They met once a month, each subcommittee involved in its own projects. The group's greatest accomplishment was a community garden that rented beds to forty different people, but most gardeners were not even aware that this was an STI project. Gaea Swinford, an STI activist and board member of Transition US, noted that Sandpoint's history as a town with a vibrant hippie subculture and strong community involvement meant: "A lot of this work has been going on for years and years, it just doesn't call itself 'Transition.'"[7]

In the United States, as Mooallem himself noted, "the American arm of the movement [was] expanding far faster than it [was] accomplishing anything."[8] While an impressive 12–22 percent of my respondents had been active in Transition Towns, most added that they had actually been to only one or two meetings. An Illinois man, for example, said, "I did go to the kickoff (er, 'Great Unveiling') of our local Transition Town group, but haven't really been involved with it," while a Colorado woman "attended one local meeting last summer": "The group did not seem very stable." A Wisconsin woman who had "moved to [a] smaller area than where [she] used to live" so as to drive less said: "I have recently joined a peak oil group, but do not attend meetings. I enjoy the listserv and discussions, though." And a Texan reported that he had attended a Transition meeting only once: "I was impressed with the purposelessness of the exercise, other than emotional catharsis."[9] One observer who had been active in Transition Towns dismissed the movement (off the record) as mere "window dressing," and others seemed to second this analysis.[10]

The vast majority of participants in the peak oil movement were well-aware that individualistic responses to impending (and current) environmental crises would be ineffective. As Richard Heinberg wrote in *The Party's Over: Oil, War and the Fate of Industrial Societies* (which three of four peakists had read): "The strategy of individualist survivalism will likely offer only temporary and uncertain refuge during the energy downslope. True individual and family security will come only with community solidarity and interdependence." Or, as a teenager from Kentucky put it, "we need to stick to community in order to prosper" during the expected peak oil crisis, instead of becoming "insular individuals." However, a number of respondents noted that the rhetoric of community response was quite different from what they themselves had witnessed. One frustrated man, an Illinois resident in his early fifties who thought that the peak oil phenomenon most closely resembled a "group of unrelated individuals following their own goals," asked: "Pray tell, where are there any real transition towns or sustainable communities? Amish is about all I've ever seen." A socialist from Oregon complained that "the 'community' talk is not matched yet by" concrete actions. While almost all respondents pointed to the need for a collective response to ecological crisis, few had real-world experience to speak of.[11]

The lack of collective action becomes apparent when we consider the number of peakists who had never physically met another believer but made major changes to their lives, such as the following:

I moved out to my land in the country, grew a garden, planted fruit trees, keeping the old car, added insulation to the house, planted shade trees, etc.

Food storage, gardening skills, tree planting.

I've terminated my studies and developed skills to allow me to be self-employed in a profession that I enjoy and which stimulates me. I've decided against having children. . . . It has affected what I look for in a partner (peak awareness).

I moved 6 years ago, researched cities with the best water quality and furthest from nuclear reactors and not on the ocean. I now own a very efficient organic farm, and have hybrid vehicles, soon to be totally electric. I have been converting to wood, installing solar panels, not purchasing gasoline appliances.

[Planted a] garden, root cellar soon, small wood lot, improving home insulation, just installed wood stove . . . building wood gassifiers [sic] for transport/backup power/homestead chores.

I gave up my car and ride a bike everywhere now. I installed a little solar, and I take steps to try to prepare for the coming squeeze.

Downsized, moved closer to urban area, bought a home with solar panels, set up food storage, kitchen garden, raising rabbits for food. Looking at farmland. Invested in physical gold and silver. Left the stock market.

Learned a traditional trade. Started a kitchen garden; learned gardening. Learned permaculture, put it in practice.[12]

These significant yet solitary preparations were not anomalies. For the most part, peakists prepped alone: almost three-quarters had begun stockpiling food, over a third had purchased a more energy-efficient car, one of four had moved to a smaller or more energy-efficient home, and one of five had changed occupation.

While comparing different historical periods is a hazardous enterprise, we should remember that considerations similar to peak everything have sparked a number of collective responses or political movements in recent American history. For example, activism around environmental and population concerns, another limit to growth that is often cited by peakists, was a major factor in the development of many communes and utopian communities in the 1960s. The environmental historian Adam Rome noted: "Especially in the countryside . . . many of the hippies were not just seeking to commune with nature. They also were motivated by apocalyptic visions of the collapse of industrial civilization."[13] In the 1970s and 1980s, fears about a different energy source, nuclear power, led tens of millions of Americans to participate in protests against the nuclear power plant industry. Apocalyptic environmentalism and a similar belief in the inability of traditional politics to achieve a necessary social and political transformation have also motivated a number of radical environmental groups, such as Earth First!.

These very different responses to similar fears in recent American history suggest the different routes that peakist concerns might have taken. Most peakist responses can be considered individualistic according to a definition of *individualism* that stresses *personalism*, which Paul Lichterman defines as a perspective that "upholds a personal self that lives with ambivalence towards, and often in tension with, the institutional or communal standards that surround it."[14] Personalism is distinct from narcissism, selfishness, and isolation in that the personalist individual does not necessarily avoid social ties or community activities but "favors and conducts a fundamentally individualized relationship to any such community."[15] (Personalist individualism also draws a sharp line around the nuclear family and reflects, as I discuss in chapter 5, a conception of the individual supported by raced and gendered notions of masculinity.) Although this type of individualism may appear to be human nature to some Americans, cross-cultural studies make clear that "the emphasis on personal development and personalized initiative" is not shared equally by all cultures.[16] Nor has it been static in the United States—communitarian authors such as Amitai Etzioni argue that this type of individualism, more conducive to therapy or personal support groups than political engagement, has been on the rise since the 1970s.[17]

The Libertarian Shift

Sociologists such as Robert Bellah and Robert Putnam have discussed the rise of individualism over the last three decades in great detail (see the introduction), but my study of the peak oil movement suggests that there is a largely ignored political dimension to the changes they observed. To understand the personalist individualism of peakists, we need to explore the rise of libertarian political ideals in the United States. While most Americans publicly identify as conservative or liberal, recent polls show that the number who would identify as libertarian has been growing steadily over the last two decades. According to a 2009 poll, for example, 59 percent of Americans described themselves as "fiscally conservative and socially liberal," which drops only to 44 percent if the position is described as "fiscally conservative and socially liberal, also known as libertarian."[18] Another survey showed that, from 2001 to 2009, the number of Americans who want less "government intervention" and less "promotion of social values from the government" had increased by roughly 35 percent. In 2011, the *New York Times* created a "libertarian index" measuring social liberalism and economic conservatism that showed a nearly 30 percent increase since 1993 in this libertarian nexus.[19] In this book, I use the term *libertarian* to encompass not only dedicated ideological libertarians but also neoliberals, the politically disaffected, and others distrustful of large organizations (such as the state), who often assume libertarian positions by default.

A discussion of libertarianism requires a brief definition and a short historical detour. The core political value for ideological libertarians is the liberty of the individual, who should be free to live his life any way he chooses, as long as he does not violate the equal rights of others. Private property is considered sacrosanct, and government's only responsibility is to protect it, prevent fraud, and ensure the successful function of free markets. Government should never impinge on the rights of the individual, especially in the removal of private property— some libertarians consider taxes that fund anything except national defense to be a form of theft.[20]

While conservatism has received a great deal of attention in recent scholarship, the focus on the conservative movement can potentially obscure the extent to which its ideas have moved beyond self-identified conservatives. The rich history of modern American conservatism, detailed in nearly one hundred monographs, paints a now-familiar picture of grassroots, white, conservative activists mobilizing in the 1950s against desegregation and communism in groups like the John Birch Society.[21] In the 1960s, they opposed liberal Great Society programs and leftist movements, and the locus of the movement moved to the South, the Sun Belt, and suburbia.[22] In the 1970s, they assumed control of the Republican Party, merged with Christian evangelicals, and seized national power with the election of Ronald Reagan in 1980.[23]

While a second wave of conservative historians is rightly beginning to question why conservatives failed to achieve many of their objectives, such as rolling back New Deal and Great Society programs, we should not forget that conservatives came to possess something approaching intellectual hegemony in the early twenty-first-century United States, and we might begin to turn our attention from the various causes of the rise of conservatism to its effects. Assessing the impact of conservative ideals requires that we define them, of course. Although the concerns of conservatives, rhetorical and substantive, explicit and implicit, have changed over time, there are a number of core aspects of conservatism that have remained relatively consistent for nearly half a century.[24] They include anticommunism, opposition to movements for racial and gender equality, a commitment to "traditional" sexual norms, and a libertarian (free market) approach to economics.[25] Of these core principles, scholars and pundits have paid the most attention to the role of social issues—perspectives on race, gender, sexuality, and "family values."[26] However, these are issues on which Americans have been actively fighting "culture wars" for over two decades, battles in which conservatives have not, on the whole, gained much ground.[27]

On the other hand, conservative principles of individual choice, reliance on the free market, and the need for a smaller state are now central subjects of debate in mainstream political discourse. These positions have (recently) been more frequently described by scholars as *neoliberal* than *libertarian*. Libertarianism and neoliberalism share the same intellectual roots in Austrian economics, both have limitless faith in markets, and both conceive of society as a collection of rational, independent, self-interested actors. On certain issues they might be understood as two sides of the same coin: while one emphasizes market fundamentalism, the other presents a philosophy of conservative individualism that is often only implied by neoliberal policies. The two ideas do have significant differences, however. On questions of foreign policy, for example, libertarians tend toward isolationism, neoliberals toward internationalism. Even on economic issues there are major differences. While economic neoliberalism claims that free market capitalism can provide growth, efficiency, and even prosperity for all, it recognizes that any market must be created and enforced by federal governments and international treaties. Neoliberalism therefore departs from libertarian orthodoxy, which views government interference in the market as fundamentally problematic—in the perfect libertarian world, free markets would simply occur naturally and require no regulation.[28] Whereas scholars have described the influence of neoliberalism around the globe, libertarianism as a political and personal philosophy is an especially strong current in American political culture.[29] This should not surprise us since, as Seymour Lipset put it in his analysis of American exceptionalism, "viewed cross-nationally, Americans are the most antistatist . . . population among the democratic nations."[30]

Even so, the libertarian conception of the individual as autonomous and detached was unpopular in the United States for much of the twentieth century, from the Great Depression until the 1970s. At midcentury, the consensus view among academics and politicians of all stripes was that the welfare state that emerged with the New Deal was a permanent improvement on the capitalist economic model since it softened capitalism's negative consequences, such as poverty and environmental degradation. In 1971, President Nixon famously announced that he too was "a Keynesian."[31] During the 1960s, however, conservative activists began to evangelize a libertarian vision of minimal government. Milton Friedman promoted libertarian ideas in best-selling books and a weekly *Newsweek* column, translating philosophy into practical policies such as school vouchers and the privatization of Social Security. These ideas moved into mainstream politics with the 1964 presidential campaign of Barry Goldwater, who inspired activists and future politicians such as Ronald Reagan. Well-heeled think tanks such as the Cato Institute, the Competitive Enterprise Institute, the Hoover Institution, and the Tax Foundation began churning out policy papers legitimizing what had previously been considered a fringe perspective.

Libertarianism is often considered a conservative political philosophy, but its emphasis on the First Amendment and the civil liberties of the individual makes it a combination of economic conservatism and social liberalism. Thus, specific libertarian ideas have intermittently appealed to leftists also. For example, in the 1970s, the counterculture adopted the libertarian rhetoric of individual free choice. The idea that social and cultural rebellion could be achieved by simply exerting individual autonomy in social matters, such as sexual identity and lifestyle (consumer) choices, paralleled the arguments that conservatives were making about economics. William E. Hudson observed that, "while the counterculture had a strong communitarian strain" that tends to dominate the public memory of the era, its "broadest political impact was more libertarian": "The individualist, more libertarian, motto to 'do your own thing' had the strongest resonance among the young."[32] In this way, libertarian thought moved closer to the vital center throughout the "Me Decade" of the 1970s.

As historians such as Daniel T. Rodgers have noted, viewing the rise of libertarian ideals in the late twentieth century as merely a triumph for the conservative movement "harbors only half the truth." Part of the remaining half, then, is that some libertarian policies were suggested first by liberals—deregulation in the late 1970s being a case in point. But, perhaps more importantly, seemingly unrelated trends in American politics and culture dovetailed with the libertarian shift. The public scandal of Watergate and Nixon's eventual resignation, for example, may have benefited Democrats in the short term, but over the long haul it set the stage for the precipitous decline of the public's trust in government at all levels, which buoyed libertarian calls for smaller government. Meanwhile, the rise of what has been termed *identity politics* on the Left in the 1970s and 1980s functioned to imaginatively divide the nation or society into what Rodgers called "little platoons of society," where "the social contract shrank imaginatively into smaller, more partial contracts: visions of smaller communities of virtue and engagement—if not communities composed simply of one rights-holding self."[33] In this context, a government that had less of the people's trust, faced with the difficulty of satisfying a heterogeneous nation with conflicting demands, lost the public battle for intellectual hegemony against a mechanism that promised a simpler, cleaner, more efficient, and apolitical solution to social, economic, and environmental problems: the market.

Although Ronald Reagan was not the libertarian that recent conservatives make him out to be (especially in foreign policy), he rhetorically promoted a libertarian agenda of individualism, deregulation, and faith in the free market. Despite the very public failures of the first wave of deregulation in the 1980s—such as escalating air travel fares and the savings and loan crisis—Congress continued to drop regulatory restrictions left and right, including limits on credit card interest, oversight of interstate banking and mortgage lending, and Glass-Steagall restrictions on commercial bank involvement in investment banking and stock brokerage. By the 1990s, the national discourse was about not the embrace of the market, which was a fait accompli, but social issues and culture wars. But, as scholars such as Steven Teles and Daniel A. Kenney have shown, social conservatism remained isolated to the United States, while libertarian think tanks "created outposts throughout the world," advocating for neoliberal solutions under the guise of apolitical technocratic management.[34]

The wave of deregulation continued under President Clinton, an exemplar of neoliberalism, who famously claimed in 1992 that "the era of big government is over."[35] In the first decade of the twenty-first century, the Tea Party succeeded in pushing libertarian ideas, such as the intention to "starve the beast" of government programs by decreasing revenue, into the American mainstream. George W. Bush relentlessly promoted the market as a central metaphor for his presidency, striving to create an "ownership society," an idea that portrayed each American as a chief executive officer, making people's very lives investment decisions. Former speaker Tip O'Neill famously referred to Social Security as the *third rail* of American politics,[36] but the campaign to privatize Social Security accounts became, in the 2010s, a public goal for Republicans. While the Occupy Wall Street movement represented one reaction to the economic recession of 2008, the Tea Party and the Republican Party called for what most observers would recognize as austerity measures, including increases in tax cuts, privatization, and reduction in government services. The contrast between the United States and similar industrialized nations has rarely been so stark: while European governments attempt to implement austerity measures in spite of massive protests against them, the United States generated populist calls for increased austerity during and after the economic recession.[37]

The libertarian shift has influenced American political culture, visions of the future, and conceptions of community, citizenship, and the self. In the last decade, a wide group of scholars has examined the relationship between "neoliberalism and everyday life," as the title of one recent collection put it.[38] Neoliberalism and libertarianism have a similar footprint on social worlds—the libertarian conception of society as merely a collection of individuals might just as well describe the assumptions behind and consequences of many neoliberal policies. Although scholars have not performed a similar exploration of the libertarian influence on the social realm, this is a potentially rich vein of criticism, especially in the United States, where the tradition of individualism has led to an easy acceptance of libertarian ideas.

Libertarianism in the Peak Oil Movement

In the influential 1985 *Habits of the Heart*, Bellah and his coauthors noted that the tendency to think of the individual as outside the framework of social institutions leaves people "without a collective context in which one might act as a participant to change the institutional structures that frustrate and limit."[39] As we shall see in the next chapter, American peakists believed that the political will or ability to change the world did not exist in the United States. In this context, the Peak Shrink's claim that "economic and planetary collapse" is actually a "therapeutic issue"—as opposed to a social issue or political issue— takes on a new resonance, reflecting the rise of a therapeutic culture that is certainly consonant with libertarian individualism.[40] As the political scientist Carl Boggs noted in his turn-of-the-millennium analysis of depoliticization, "whereas popular movements (civil rights, feminism, ecology, and so on) generally affirm collective values and action, the therapeutic shift tends to focus on the self in a way that detaches individuals from their social context" and, thus, leaves them powerless.[41] In such a position, all that the individual citizen can do is alter his or her perspective and address collective problems with psychology instead of politics. Indeed, the Peak Shrink informed her readers: "[It is] not your job to fight [the 'Panglossian disorder']. You cannot. . . . Maybe this year, maybe next, you may be thought of as the sane one after all."[42]

The personalist individualism that is characteristic of the libertarian shift in the United States was evident in my surveys. American peakists displayed a consistently more individualistic outlook than did adherents from other countries. When asked about political or other affiliations, respondents often described themselves as individuals in flux, moving between different local, religious, political, or virtual communities while never being fully committed to any of them. Unlike participants in other movements, peakists seemed to view themselves as merely a collection of like-minded individuals.

Even so, the political skepticism of peakists, which is of a piece with skepticism among most Americans, is uncommon. Although one of the typical characteristics of Americans on the Left is faith in the ability of the federal government to solve social problems, peakists betrayed the antistatist foundation of contemporary American politics in their distrust of the government. In April 2010, 21 percent of Americans said that most members of Congress deserved reelection; responding less than a year later to the same question, the number for peakists was 3 percent.[43] This degree of skepticism of the ability of the federal government to manage major societal issues—whether ecological crises, health care, or desegregation—is considered one of the characteristics of conservatives, not what one expects from a group composed of self-identified progressives and socialists. It is indicative of a tectonic realignment in American political culture that transcends party politics.

Liberal Survivalism

The individualism of peak oil believers combined two orientations that have infrequently intersected in recent American history: liberalism and survivalism. Since the 1980s, the term *survivalist* has generally referred to groups of conservative, "marginalized, poorly-educated, lower-class white males" that had retreated from the public sphere, such as militias, the Aryan Nation, and the Posse Comitatus.[44] Although the peakist demographic was quite different, these two assemblages have their similarities, and many of the claims that the sociologist Richard J. Mitchell made about survivalists in the 1990s would hold true for peakists a decade later. Survivalism, he wrote, "finds its niche in dramatic doubt, in a rhetoric of radical skepticism toward the prospects of contemporary institutional orders." Believers create "discourses of pending need, speculative circumstances of crisis and concern wherein major social institutions face imminent serious erosion or total dissolution and in which survivalists themselves play central roles in reprioritized revisioning, recovery, and renewal."[45] Perhaps recognizing these similarities, 38 percent of peakists agreed with the statement, "Most responses to peak oil are a form of survivalism," but went to great lengths to differentiate themselves from "the guns-and-ammo survivalists that you usually think of."[46] In my surveys, there was some truth to the claim that survivalism was practiced more by those on the Right—while most American peakists practiced some level of survival preparation, self-identified conservatives were more likely to have stockpiled food and other supplies than self-identified liberals.[47]

The closest corollary to the peakist union of liberalism and survivalism in recent American history is the "back-to-the-land" movement. Throughout the 1970s and 1980s, almost one million urban Americans relocated to rural homesteads to practice self-sufficiency.[48] Like peakists, most were white leftists who adopted a voluntary simplicity as a result of their radical beliefs. Back-to-the-landers even had similar motivations, with many citing environmental issues and overpopulation as a major factor in their decision to relocate; in one survey, 38 percent claimed that one of the reasons they had moved was that they "feared an economic collapse."[49] However, back-to-the-landers were not motivated by a vision of secular apocalypse, and unlike most survivalists and peakists they were not prepping to survive an imminent crisis. As Jeffrey Jacob noted: "The back-to-the-land movement was, in its own quiet way, a broad-based protest against what the spirit of the sixties saw as the irrational materialism of urban life."[50] Back-to-the-landers also tended to view their actions in a political light, even if they were only setting an example for others and not actively engaging in formal or traditional political action themselves.

The peak oil movement, like the back-to-the-land movement, is a reflection of its historical moment. The back-to-the-land movement demonstrated the breadth of environmental politics that grew throughout the 1970s. Not only were Americans engaged en masse in activist groups such as Greenpeace, but, in keeping with the maxim that "the personal is political," they also transformed their personal lives to express their political and environmental beliefs. Similarly, the advent of liberal survivalism in the twenty-first century required two ingredients: perceived apocalyptic threats, such as resource depletion and climate change, and an individualism characteristic of conservative survivalists to become widespread enough for those on the Left to turn to solitary survivalism. This individualism was partially encouraged, paradoxically, by the primary social technology of our age: the Internet.

Networked Individualism

If one factor in the individualism of the peak oil phenomenon was the broad-based libertarian shift, the second, related factor is just as far-reaching: the effect of Internet technology on participants in what was primarily a virtual movement. To explore the influence of the Internet on the peak oil movement, I take a necessary (and brief) detour through the libertarian origins of the Internet before discussing the "network effect," the influence of Internet technologies on their users.

The history of the Internet plainly reveals its connection to libertarian ideals. First, in the 1950s, a number of early developers saw the personal computer as a technology of self-development, a prosthesis that might "augment the human intellect."[51] Later, the interconnected network of global computers that became known as the Internet was developed in the 1960s and 1970s by US army engineers as an open, minimalist, neutral, and decentralized system. This was partially because it was designed to survive a possible nuclear strike by the Soviet Union that might destroy major military bases, but this original architecture also picked up the traces of the libertarian leanings of many early computer programmers.[52] These architects and early users "envisioned cyberspace as a new, libertarian post-national social order," and they created a network that would enable local control over the medium, instead of a centrally organized, government-run network.[53] This openness allowed each user to alter the network itself, which was a revolutionary idea at the time. In this way, as Jack Goldsmith and Tim Wu put it, one can find "strains of American libertarianism, and even 1960s idealism," in "the universal language of the Internet."[54] Indeed, as Fred Turner has shown, there is a great deal of continuity between the liberation methodologies of the 1960s counterculture (as distinct from the New Left) and the digital utopianism of the 1980s and 1990s, and major countercultural figures (such as Stewart Brand) were instrumental in popularizing key concepts related to personal computing and the Internet.[55]

A second convergence between computer technologies and a different strand of libertarian ideals occurred in the 1980s, when the "Information Revolution" of personal computers and other communication technologies enjoyed a synergistic relationship with the libertarian rhetoric of President Reagan and other movement conservatives. As the historian Philip Jenkins noted: "The flourishing technology industries—especially computing—looked like wonderful commercials for decentralization and deregulation, entrepreneurship and risk taking, heroic private enterprise and free trade."[56] Economic libertarians could point to companies such as Apple and Microsoft as evidence that their policies were spurring innovation and economic growth. The influence of libertarian-tinged computer culture was widespread—for example, only in the context of what observers began to call *cyberlibertarianism* could computer hackers be seen not as high-tech criminals but as heroic individuals standing up to an overbearing state.[57] Although "not everyone who bought a computer came to believe in the free market," of course, the libertarian mythology of the personal computer could serve as a self-fulfilling prophecy of the network effect. As Thomas Streeter wrote: "The experience of reading about, buying, and using microcomputers created a kind of congruence between an everyday life experience and the neoclassical economic vision—the vision of a world of isolated individuals operating apart, without dependence on others, individuals in a condition of self-mastery, rationally calculating prices and technology."[58] In the 1980s, this romantic vision of the liberatory potential of personal computing was articulated to the ascendant philosophy of libertarian individualism.

By the time the Internet as we know it fully emerged in the late 1990s, it had also been burnished in the fire of neoliberalism. In the 1990s, the Internet faced competition from an alternative, proprietary mode of connection that is now often forgotten. Companies such as

America Online (AOL), Prodigy, and Compuserve provided access to the World Wide Web but did not interconnect, and a user on one service could not contact someone on another. These services offered a very different experience: in contrast to the open content of the Internet, each decided how to filter content and what sort of chat rooms to allow. Users had only partial anonymity and much less control over each network. The proprietary model was eventually overwhelmed by the Internet, a civil libertarian's dream with almost no filters for values or content.

The libertarian direction of the online experience, which now seems completely natural and beyond politics, reflected the neoliberal zeitgeist of the 1990s. The Clinton administration ensured that the Internet would remain mostly unregulated by deregulating telecommunications services and information technology products. For example, in 1997 the United States signed three major agreements with the World Trade Organization that limited regulation of the Internet. On nearly all sides of the debates on the direction of Internet communications in the early 1990s, the rhetoric of free markets held sway. For politicians, the exciting new technology "was all about unleashing entrepreneurs" and "promoting innovation."[59] In 1997, when the Clinton administration released its Framework for Global Electronic Commerce, it recommended that "the Internet should develop as a market driven arena" instead of "a regulated industry."[60]

The tech industry, with its continuing libertarian streak, recognized that Internet policies of openness, decentralization, and individualism could be slotted into neoliberal policies. Its primary mouthpiece, the magazine *Wired* (edited by Kevin Kelly, the former editor and publisher of the *Whole Earth Catalog*) promoted an ideology that Paulina Borsook has described as "technolibertarian," even placing Newt Gingrich on its cover.[61] The software mogul Bill Gates, in the influential *The Road Ahead* (1995), predicted that the Internet would finally deliver on the promise of "friction-free capitalism," and a 1995 essay by Christopher Anderson claimed: "The growth of the Net is not a fluke or a fad, but the consequence of unleashing the power of individual creativity. If it were an economy, it would be the triumph of the free market over central planning."[62] The notable lack of debate about the direction of the Internet at a crucial moment in its development was rightly attributed by John Bellamy Foster and Robert W. McChesney to "the digital revolution exploding at precisely the moment that neoliberalism was in ascendance, its flowery rhetoric concerning 'free markets' most redolent."[63] As Turner summarized: "By the end of the [1990s], the libertarian, utopian, populist depiction of the Internet could be heard echoing in the halls of Congress, the board rooms of Fortune 500 corporations, the chat rooms of cyberspace, and the kitchens and living rooms of individual American investors."[64]

The Network Effect

That the Internet would have an influence on its users—which would soon include most Americans—was assumed by enthusiasts such as Gates and Anderson as well as early scholars of the technology. In the first wave of scholarship in the 1990s, a number of prominent authors, such as Sherry Turkle, Robert Putnam, and Robert Kraut, argued that time spent online might isolate users from the real world and, thereby, weaken commitments to civic, political, and public life. The Internet, some feared, would encourage social isolation and lead to the further decline of communities of place.[65]

The second wave of criticism, which lasted until late in the period 2000–2010, overturned this initial concern. Most early twenty-first-century media commentary and scholarship about the Internet was characterized by a fascination with Web 2.0 technologies, such as social media and blogs, and the changes to work, politics, and social life that they effected. In a 2009 meta-analysis of cross-disciplinary answers to the question of the network effect on political and civic engagement, Shelley Boulianne found that thirty-three of the thirty-eight studies she examined reported "positive" effects of the Internet on its users.[66] Although the Internet was commonly blamed for a decline in attention span and provoked an intermittent moral panic around sexual predation and cyberbullying, the media has, for the most part, "herald[ed] a new golden age of access and participation" via each new network development, such as high-speed Internet, WiFi, Facebook, Twitter, and smartphones.[67]

At the same time, scholars in sociology and related disciplines pushed back at attempts to measure the network effect at all by emphasizing the role of "social construction" in the study of technology.[68] Social construction theory claims that technologies do not in themselves have any effects on their users; instead, we should investigate the way that technologies are shaped by gender, race, class, culture, and other factors. Although social construction was a generative concept when it first appeared in the mid-1980s (and still is), by the first decade of the twenty-first century the concept was often being used to foreclose on potential scholarship.[69] Some social construction scholars caricatured their antagonists as "technological determinists" who believed that technology was the only cause of historical change. As Robert Hassan pointed out, "technological determinism" was considered "something unmentionable, a kind of proscribed thought crime, a dark and hidden 'undercurrent' that occasionally 'betrays' itself in the careless writings of some social theorists."[70]

As such, scholarship was actually lagging behind commonsense recognition of the Internet's influence. As Hassan himself noted: "At the intuitive level at least, we recognize that our becoming as one with the network society is occurring rapidly. We are everyday being offered (and are greedily accepting) deepening connections with a shrinking networked world."[71] Indeed, this seemingly unquestioned acceptance of new technologies, which quickly become expected of every functioning member of society, has been labeled *technological somnambulism*.[72] Paradoxically, then, as the Internet became deeply embedded in almost every aspect of American culture, it became more difficult for scholars to write about it. This is in part because many failed to identify the Internet in its proper historical context—not as a series of individual new media (e.g., Facebook as a stand-alone subject) but as a system, a network of technologies that includes personal computing, the World Wide Web, social media, and smartphones.[73] These are important not as isolated developments but as aspects of a metatechnology that has created its own context, a system that now constitutes, as the sociologist Manuel Castells put it, "the social morphology of our societies."[74]

The answer to the question of the Internet's political and social influence surely lies somewhere in between the poles of social construction and technological determinism. One must acknowledge the central roles of innovators, Web site developers, forum moderators, and individual users. But as Evgeny Morozov has argued: "If technological determinism is dangerous, so is its opposite: a bland refusal to see that certain technologies, by their very constitution, are more likely to produce certain social and political outcomes than other technologies, once embedded into enabling social environments."[75] Although its potential for group activity has been highly touted, we should recall that the Internet is a fundamentally individualistic medium. The conceptual stalemate between technological determinism and social construction can be fruitfully sidestepped by applying the concept of social affordances, which is used by scholars of technology to indicate the preferred or intended use of technologies. For example, while a bench can be used for a variety of purposes, it has an affordance for sitting.[76] The basic physical act of accessing the Internet is worth mentioning here: alone at her computer, "it is the individual, and not the household or the group, that is the primary unit of connectivity."[77] Since personal computers and the Internet are used and accessed alone, by a solitary individual, they have an affordance for individualism, although they can be and are used for a variety of purposes.

This affordance can be highlighted by comparing the Internet with two other twentieth-century technologies that also had powerful impacts on their users: radio and television. Both can be and often are experienced by groups of people simultaneously, such as the family in its living room listening to or viewing the same program. In this way, they have a very different social affordance. These technologies do not have the same potential for self-expression as the Internet since they construct identity vertically, from one (the television show's creators) to many (the audience). As such, they are indeed "mass media capable of constructing mass audiences and mass consciousness."[78] As Ima Tubella has argued, the Internet is fundamentally different from these technologies; instead of a mass or collective identity, it "influences the construction of individual identity, as individuals increasingly rely on their own resources to construct a coherent identity for themselves in an open process of self formation," by selecting their applications and unique browsing path, posting and commenting, and creating or altering Web pages.[79]

The basic fact of the Internet's social affordance for networked individualism, rarely acknowledged by scholars wary of being labeled technological determinists, was recognized at some level by my respondents, who cited the movement's reliance on the network as the most frequent explanation for its lack of collective identity. An Alabama man in his forties said that the peak oil movement must move "beyond an electronic medium and a few meetings once every blue moon," and an Ohio civil service employee claimed that "the nature of the Internet" means that the virtual peak oil movement is "not 'real.'" Another respondent, in his early thirties, simply noted that "community means something different on the Internet," while an educator in his fifties specified that the reason some peakists have not formed communities is because they are "conditioned by the greater culture to behave as autonomous individuals."[80] While the Internet can be and is used to create and enhance communities, it has a social affordance for personalist individualism.

Peak Oil Online

As documented in the last chapter, digital technologies enabled the very existence of peak oil as a mass phenomenon, but they also shaped the ways that believers conceived of and acted on their newfound ideology. While the Internet has been praised for its ability to create new communities and enable political action, the example of the peak oil movement shows that in some cases this new medium serves to nudge its users in an individualistic direction, reflecting the libertarian origins and design of the technology itself. On the Internet, we are not citizens or community members, defined by our relationship to a place or group, but isolated "users," increasing the tendency to imagine society as made up of abstract individuals pursuing their rational interests in a marketplace. Participants in the peakist virtual community formed strong but often short-lived ties to each other, and the paucity of face-to-face meetings had a powerful effect on this group's struggle to form a sense of collective identity.

For the most part, the peak oil movement functioned as a virtual community, which can be defined as a group of "people with shared interests or goals for whom electronic communication is a primary form of interaction."[81] While some peakists congregated at conferences and lectures, most communicated primarily, if not solely, through the Internet. Online, they created the rich virtual subculture described in chapter 1, with Web sites, blogs, podcasts, YouTube channels, graphic novels, video games, musicals, and online forums dedicated to resource depletion, climate change, and collapse. Although some of these artifacts were intended to evangelize the peak oil theory to the uninitiated, they had the most influence on hard-core believers. As Kris Can said of her video blog, which included entries like "Postcard from the Future" and "Inserting Peak Oil into the Conversation": "Lots of people e-mail me about my show and discuss the different topics that I bring up. I originally thought, wow, what am I doing, I'm just preaching to the choir. And then someone said, the choir needs community too, and I thought, yes, absolutely."[82]

Most scholars have looked at virtual communities that are formed and maintained on a single Web site, in part because this is simply easier to do, but we have seen that the peakist community was spread over a network of locations. Seventy percent of American peakists regularly visited at least three peak oil Web sites, and 29 percent regularly checked at least five.[83] Some of these have already been mentioned, but there were dozens of others that respondents brought to my attention, such as The Oil Age, Michael Ruppert's Collapse Network, The Archdruid Report ("Druid perspectives on nature, culture, and the future of industrial society"), Peak Oil Crisis, Decline of the Empire, The Post-Carbon Institute, Malthusia, Silent Country, Peak Moment Television ("Locally reliant living for challenging times"), and Early Warning: Risks to Global Civilization.[84]

To explore the kinds of discussion and community that these sites provide, we might inspect Peak Oil News and Message Boards (www .peakoil.com). In its first eight years online (2004–11), the 30,000 members of Peak Oil News wrote over 950,000 posts on 26,000 different topics. Over 2,000 members contributed at least 20 posts to the site. Peakists tended to see the primary goal of their participation as information gathering, and the busiest forums reflected their interest in facts and figures: "Economics and Finance" had over 2,700 subtopics and 100,000 posts, with "Energy Technology," "Environment," and "Planning for the Future" close behind. Common subjects in these respective forums were the impact of resource depletion on the global economy, technological developments in renewable energy efficiency, climate change, and practical steps to prepare for the coming collapse. Scattered throughout the site are frequent threads on the health and tenor of Peak Oil News itself, reflecting a general interest in the state of the virtual commons.

The popularity of the "Open Topic Discussion," with over 9,000 subtopics and 270,000 total posts, shows that adherents were not solely interested in gathering information but sought guidance from and provided advice to strangers with whom they shared a deep connection. Friendships were formed—one of three adherents who regularly checked at least three peakist Web sites reported having gained friends as a result of their activities online.[85] The popularity of the "Medical Issues Forum," which also dealt with psychological issues, reflects the affective investment of many believers in the digital community. The most popular thread there, with over 45,000 comments, was "My Doom-O-Meter is jittering towards max," in which posters discussed the common experience of depression and even suicidal ideation that, they said, they were often afraid to confess to their friends and family. (Because of the emotional impact of peakism, this subject was so common that site administrators created a separate thread, "Don't Do It," reminding members: "Refrain from endorsing suicide. Its explicitly against the [Code of Conduct].")[86] Fifty-five percent of survey respondents reported that one of the benefits of visiting peak oil Web sites was the emotional "comfort of knowing that other people have the same thoughts." This fellowship was evident throughout the online forums, although it was typically expressed indirectly, in gallows humor. In the "Doom-O-Meter" thread, nefor example, one member responded to a particularly apocalyptic series of comments by writing, "I know I can always count on you all to cheer me up," and then added: "Hoping I've got 2–3 years. . . . Stay calm & collected. Navigate. Trust in your preparations. Have faith. If you don't despair, you'll make it." The original poster, "Heineken," wrote back: "Just terrific comments from a terrific bunch of people I sure as hell wish were my neighbors."[87]

Unfortunately for Heineken, who had recently moved to rural Virginia, they were not his neighbors. Unlike traditional communities of place, such as neighborhood organizations or sites of worship, peakists who connected online were usually geographically distant from one another. Many were thus unable to meet other believers and create the collective or communitarian response to the perceived challenge that they desired. A survey respondent in his late forties in Arizona wrote, with some despair, that he lived in a "flyover town" in "the middle of nowhere," while a "liberal" septuagenarian complained that, since "too few people are aware of peak oil" in his small town, "the only place we meet is on the net," and many others lamented that there were no local groups in their area.[88] Although scholars continue to debate the emotional, psychological, and political importance of communities formed over the Internet, the difficulty of transitioning to real-world meetings that result in stronger, longer-lasting ties is widely agreed on.[89] In this literature, the strength or weakness of a connection or tie is defined as a "combination of the amount of time, the emotional intensity, the intimacy (mutual confiding) and reciprocal services which characterize the tie."[90] While the peak oil online community provided ties that were quite strong, they were generally fleeting and rarely transitioned into real-world friendships.

Although most of my respondents participated in online forums, not one reported having met these interlocutors in person. This is the logical consequence of two central features of virtual communities: voluntary participation and anonymity. Friendships or connections developed online are dependent on each participant voluntary returning to the same forum and publicly participating. Anonymity, which provides a sense of security and freedom of debate, also makes it difficult for members to continue their connections offline unless they have specifically indicated such a desire and shared contact information, a situation that appeared to be rare. For example, in 2008 one heavy commenter, "PenultimateManStanding," created a thread called "I'd Like to Meet All of You" in which he expressed his desire to do just that. After a series of responses expressed some fear that forum participants would be different in "real life" from their online personas, he responded in exasperation: "Well, if nobody takes this seriously, that's understandable. But I'm not kidding, I'd buy a ticket to meet many of you."[91] Such a meeting does not seem to have occurred, and PenultimateManStanding soon stopped posting entirely. Like many others, this virtual community provided conditional ties of short duration.[92] Daniel T. Rodgers described the libertarian vision of society that emerged in the 1980s as "more voluntaristic, fractured, easier to exit, and more guarded from others" than traditional American communities, but it is only in virtual communities that this vision came to its full fruition.[93]

Although a number of scholars have rightly argued that virtual communities can revive or even replace communities of place, they often analyze virtual communities that are actually grounded in existing, real-world connections and/or geographic proximity, such as Facebook.[94] When asked to define the peak oil phenomenon, peakists themselves had no trouble distinguishing it from traditional communities. Only 12 percent of respondents considered *community* an appropriate term for the movement, while 50 percent thought of peak oil as an "information hub," and 20 percent considered it to be merely "a group of unrelated individuals following their own goals." Only 9 percent considered the peak oil phenomenon a *movement* at all, reflecting a traditional conception of the term, one based on examples such as the civil rights movement. Some reasoned that peak oil lacked leaders, although there were, in fact, a number of well-known writers and speakers, such as Kunstler, Colin Campbell, Matthew Simmons, and Richard Heinberg. Others pointed to the lack of the political lobby that often defines a movement. One commenter on Peak Oil News, responding to a 2011 thread titled "What Happened to the Peak Oil Movement," asked rhetorically: "How can Peak Oil be a movement without marches & protests?"[95]

This is a generative question, although perhaps not in the way that its author intended. More vital to a social movement than hierarchies, formal structure, or a political lobby is the collective identity and sense of empowerment that often results from large group meetings, whether they are informal gatherings or political events. Scholars of social movements and crowd psychology alike have found that social movements often begin when people physically meet and see themselves as a collective, with shared interests and the potential to effect change. For example, a protester interviewed by the sociologist Jon Drury reflected on the experience of empowerment at an antiwar protest in 2001: "I think because of the amount of people there that certainly had a psychological effect and you think yeah, you know, this is going somewhere this is moving and this is going to build up momentum and get larger and larger, and because of that this is going to turn into something really positive." Another protester, at an environmental action in the 1990s, reflected on his experience in a similar way: "It was empowering in the sense that after that I had a lot of energy and, you know, the feeling after that 'Wow, you know, we can do anything.'"[96] Historians of social movements have noted, time and time again, that collective actions—whether they are meetings, protests, or even riots— can have powerful effects on participants' sense of collective identity, empowerment, and political outlook.

Individualism Online

However, this process is rarely achieved on even the busiest Internet forums. Although scholars of social movements recognize the importance of collective, physical meetings, those who study online communities have often failed to appreciate the significance of their absence on Internet groups and virtual communities. Various aspects of the online experience have been compared to a crowd (e.g., "crowdsourcing" and "crowdfunding"), but the experience of surfing the Internet and that of participating in a physical crowd have very different psychological and emotional effects. As the social psychologist Tom Postmes put it: "On a continuum of social contexts ranging from individualistic to collectivistic, the Internet and the crowd are somewhere at the extremes."[97]

Even when Web sites such as Peak Oil News publicize their membership statistics, they typically list this information in small type at the bottom of the page. If collective events such as protests provide a heightened sense of collective identification, Internet forums can have the opposite effect, partially because the only people whose presence is noted are those who actively post—so-called lurkers, who visit membership sites but do not post and are, thus, not visible, make up as much as 90 percent of traffic on some membership Web sites.[98] This dwarfing effect was reflected in the number of survey respondents who seemed to believe that only a handful of people were interested in the issue of oil depletion. For example, a man from Iowa said that there was "only a small group of believers," while a Minnesota woman in her late fifties said that "Peak Oilers are too scattered to be a real community and too few to be a movement."[99] Others suggested that there were only a few hundred people interested in the subject. The sense of collective identity that can be advanced by physical, face-to-face meetings was, thus, never achieved for participants in this virtual community.

The lack of real-world meetings thus led to a misperception of the number of participants, a misperception that is directly tied to the sense of political fatalism described in the introduction and explored in greater detail in the next chapter. Each individual decision to forgo political action was based, one could argue, on a rational calculation of the unpopularity of the concept of peak oil, our total immersion in the fossil fuel regime, the obstinacy of a culture of optimism, and the failure of recent environmental initiatives in the United States, but many social movements begin by introducing ideas that are unknown or unpopular. As John Drury and Steve Reicher have argued, political "action depends upon a calculation that one is able to overcome the forces that protect the status quo," and one of "the factors going into such a calculation include[s] the size" of the group itself.[100] If one has the sense that the group is composed of a few hundred people instead of a few hundred thousand, one will severely underestimate the group's potential "collective self-efficacy."[101] As a result of their participation in Internet forums instead of more traditional group events, such as mass meetings, rallies, or marches, peakists underestimated the size of their constituency, a fact that contributed to their sense of political inefficacy and fatalism.

Collective Challenges, Individual Solutions

Scholars of Internet psychology and recent American politics are engaged in similar conversations, but few have connected the Internet's potential to be "individuating and atomizing" to the depoliticizing tendency of increasing individualism, which has been the subject of a great deal of work by scholars of neoliberalism.[102] By assigning to the individual "relations of obligation" "solidarity and concern that were once considered part of the common good," libertarian philosophy and policies, whether packaged as neoliberalism or Tea Party Republicanism, imagine the collective as merely a collection of individuals. The purest distillation of this view was the philosopher Robert Nozick's claim that "there are only individual people, different individual people, with their own individual lives . . . [n]othing more," which Margaret Thatcher famously articulated as: "There is no society, there are individual men and women."[103] According to libertarian and neoliberal theories, societal problems are best solved by individual action in the private sphere, not government initiatives. In the United States, this perspective found a historical foothold in the tradition and mythology of rugged individualism and self-sufficiency.

However, the trend toward smaller government and privatization tends to depoliticize important social, political, and environmental issues that in the contemporary globalized, interdependent world are often too large in scope to actually be dealt with by individuals. As Henry A. Giroux has argued: "As the social became individualized [in the 1980s and 1990s], uncertainty and fear worked to depoliticize a population that is educated to believe that social problems can only be addressed through private solutions. Within such a climate, shared responsibilities gave way to shared trepidation."[104] The only way that complex global problems such as resource depletion and climate change can conceivably be addressed is through national and international bodies, not by isolated individuals. By moving the locus of responsibility from government to the individual, the shift toward libertarian ideals effectively depoliticizes major social problems and, thereby, cripples our ability to address them successfully.

Tracing the parallel and interrelated histories of libertarian thought and Internet effects helps us understand how the Internet has helped create the society that libertarian theorists had previously only imagined.[105] While one might analyze the personalist individualism of the peak oil phenomenon in the context of either the libertarian shift or the influence of Internet technology, these two developments are best understood as having a synergistic relationship. Although the Internet has a social affordance for individualism, the cyberlibertarian conception of Internet crowds as merely a collection of rational, isolated individuals is particularly powerful when political theorists and politicians are describing society in this same vein, in the "enabling discourse" of libertarian ideals.

Without the Internet, the peak oil phenomenon would not have occurred. The possibility of impending resource depletion would probably have remained where it was in the 1990s, when only a few geologists and environmentalists were concerned about these questions. But, without the Internet as its medium, these concerns might have taken a very different form, by sparking a social movement demanding that the United States transition away from reliance on nonrenewable energy sources, for example, instead of the mostly individualistic responses we saw in chapter 1. This is a counterfactual speculation but one worth considering. As a result of the libertarian shift in American political culture and the atomizing impact of the Internet, peakists primarily responded to the threat of environmental crisis as isolated individuals, even as they acknowledged that a collective response was needed.

These interrelated trends have implications far beyond the peak oil movement, as I explore in the conclusion. First, in the next chapter, I further examine the political quiescence of peakists by investigating the relationship between apocalyptic popular culture, environmental politics, and political fatalism.

FOUR

Apocalyptic Popular Culture and Political Quiescence

Some say it's all going to be *The Road*. Or *Mad Max*. Or *World Made by Hand*. WHITE FEMALE, EDITOR OF A TRANSITION VOICE MAGAZINE, VIRGINIA, FORTY-ONE TO FORTY-FIVE

Just because [peakists] have been influenced by fictional portrayals doesn't mean those scenarios are unlikely. WHITE FEMALE, PREPPER, CALIFORNIA, FIFTY-ONE TO FIFTY-FIVE

I think our American way of life is coming to an end. WHITE MALE, VIRGINIA, FIFTY-ONE TO FIFTY-FIVE

This chapter situates the peak oil phenomenon in the context of American apocalypticism and explores the relationship between environmental political engagement and narratives of regeneration through crisis in fiction, film, and the political imaginary. The coalescence of the peak oil movement was the result of a number of technological, economic, political, and environmental developments (described in chapter 2), but it also fit into the long history of apocalyptic imagination in American history. Millennial ideologies and groups are remarkably popular in the United States, and apocalypticism was a major theme in popular culture throughout the twentieth cen-

This chapter's epigraphs are taken from respondents 9592381, 9577250, 9569301.

tury. I document the growth of one form of apocalyptic popular culture, eco-apocalyptic disaster movies from the period 1990–2010, and describe its influence on peak oil believers by analyzing popular peak oil novels.

My survey respondents readily acknowledged the influence of apocalyptic popular culture on their conception of the future, and their prophecies bore a great deal of similarities to Hollywood films such as *The Postman* (Kevin Costner, 1997) and *The Day After Tomorrow* (Roland Emmerich, 2004). However, disaster films and peak oil fiction serve fundamentally different purposes for their viewers and readers. While blockbuster films offer spectacular scenes of destruction, they typically demonstrate a symbolic national unification through the transcendence of persistent divisions of race, gender, class, religion, and politics. At a historical moment when radical change through social movements or electoral politics seems unlikely to many Americans, the prophecy of national regeneration through crisis provides a means of imagining a significantly different world. This is not necessarily a utopian future but one in which Americans are shown to hold true to the ideals of their mythology: hardworking, egalitarian, community-minded and inclusive.

In peak oil fiction, such as James Howard Kunstler's widely read novel *World Made by Hand* (2008), this national mythology was challenged and rejected. Peakists saw oil depletion as a historical event that might finally bring about a revolutionary transformation and put an end to American imperialism and even capitalism. This change would be achieved not by a social movement but by the petroleum-powered American way of life tripping over its ecological limits. Motivated by anticapitalist politics, the peakist vision of the future depicts an apocalyptic reckoning with revolutionary consequences, but it also (counterintuitively) contributed to the political quiescence described in the last chapter.

The Apocalypse in American History

Apocalypse, which the folklorist Daniel Wojcik describes as "the catastrophic destruction of the world or current society, whether attributed to supernatural forces, natural forces, or human actions," has been an important theme throughout American history.[1] While peakism may strike some readers as an unusual belief system, the peak oil movement does not seem quite as fringe when situated in the context of American apocalypticism. In 1999, for example, 36 percent of Americans admitted to planning to "stockpile food and water" in preparation for the fallout of the "Y2K" computer bug, while a 2006 poll found that one-quarter of Americans believed that Jesus Christ would return to the Earth the following year.[2] Connecting contemporary events to millennial prophecies is also not uncommon—in 2002, for example, one in four Americans claimed that the Bible had predicted the 9/11 attacks.[3] While a fascination with the end times may be universal, my surveys of peakists confirmed the particular prominence of apocalyptic belief and action in the United States. Among socioeconomically similar respondents from the United States and comparably industrialized nations (Canada and the countries of Western Europe), for example, American peakists were much more likely to have stockpiled food and other supplies.[4]

These attitudes have a unique historical precedent in the United States. Early European explorers and Puritans often cast the United States as a "new Eden" burdened with millennial expectations, and the Shakers, Millerites, Seventh-Day Adventists, Mormons, Jehovah's Witnesses, and other groups developed chiliastic Christian theologies.[5] The apocalyptic currents in American religion became more conspicuous in the second half of the twentieth century as the membership rolls of evangelical Christian denominations swelled. Premillennial dispensationalism, a theology of biblical literalism that sees the return of Christ and the millennium as imminent events, moved from the fringe of American Christianity to the mainstream.[6] This shift was highlighted by the visibility of premillennial evangelists such as Billy Graham, Jerry Falwell, and Pat Robertson and the public religiosity of Ronald Reagan and George W. Bush.[7] Beyond Christianity, apocalypticism was a core element of scores of new religious movements that appeared in the post–World War II period, including various New Age groups.

While apocalypticism has typically been the purview of religion, belief in the end of the world constituted an increasingly important aspect of American secular culture during the nineteenth and twentieth centuries. Millennial ideas have been instrumental in American military conflicts (such as the Civil War), social movements (such as abolitionism, the temperance movement, and the Ghost Dance movement), and some of the central components of national mythology, including American exceptionalism and Manifest Destiny.[8]

The peak oil movement was placed in this tradition by a number of observers, described as the *liberal Left Behind* by some and derided as a *cult* by others.[9] As the narratives of conversion recounted in chapter 1 suggest, peak oil certainly bears a family resemblance to social phenomena widely acknowledged to be religious. After becoming peak oil aware, adherents shared a belief in worldly destruction and a new conception of proper or ethical conduct, they practiced rituals of prediction, prophecy, and preparation, and they often attempted to convert their friends and family.[10] Both skeptics and believers have considered the peak oil ideology to be similar to a religion. While peakism lacks a conception of the sacred or supernatural, it certainly has religious dimensions.[11] However we categorize peakism—as an ideology, a religion, or a form of radical environmentalism—it is clear that the peak oil ideology drew on various strands of millennial theology.

The Post–Cold War Boom in Doom

During the Cold War, the prospect of nuclear holocaust influenced any number of political and social developments, including, as we saw in chapter 2, the rise of modern environmentalism.[12] After the end of the Cold War and the threat of nuclear annihilation, the background hum of apocalypticism in American culture assumed new forms. Some of this activity was due to the anticipation of the year 2000 itself, which generated countless History Channel specials on apocalyptic themes as well as anxieties such as the Y2K computer bug.[13] In the early years of the twenty-first century, President George W. Bush spoke openly about his beliefs as a born-again evangelical, and the 9/11 attacks provoked a millennial paranoia throughout the country. The British *Daily Mail* summed up this mood on September 12, 2001, when it ran the headline "Apocalypse!" above an image of the crumbling World Trade Center. Many Americans apparently agreed—in the weeks after 9/11, sales of Christian prophecy literature increased by 71 percent.[14]

Long a staple of science fiction, post-apocalyptic visions moved into the mainstream of American popular culture in the twenty-first century.[15] The evangelical *Left Behind* series (1995–2007), which depicts a premillennial dispensational Christian eschatology, including the rapture of observant Christians and the rise of the Antichrist, was suddenly given, as Douglas Kellner noted, "a new cultural resonance post-9/11" and continued its remarkable run of popularity.[16] Novels such as Margaret Atwood's *Oryx and Crake* (2003) and S. M. Stirling's *Dies the Fire* (2003) represented the biggest wave of end times literature since the 1960s, and the genre received a stamp of elite literary approval in

2007 when Cormac McCarthy was awarded the Pulitzer Prize for fiction for his novel *The Road* (2006). Television shows such as *Battlestar Galactica* (2004–9), *Jericho* (2006–8), and *Falling Skies* (2011–) brought the end of the world to the small screen, while teenagers and young adults immersed themselves in post-apocalyptic landscapes in video games such as the *Fallout* (1997–2010) and *Half Life* (2001–4) series. Zombie and vampire narratives, which often threatened the complete extermination of human beings, appeared in all these formats.

Of all media platforms and genres, Hollywood disaster films exerted perhaps the strongest influence on peak oil believers and Americans in general. The disaster genre was remarkably popular in the 1970s, with hits such as *The Poseidon Adventure* (Ronald Neame, 1972) and *The Towering Inferno* (John Guillermin, 1974), but these films tended to focus on relatively small-scale crises, often isolated to a single ship, building, or airplane. When the genre returned to prominence in 1996 with *Independence Day*, it conceived of disaster on an unprecedented scale. From 1995 to 1998, there were at least three major (high-grossing) disaster movies per year, including *Independence Day* (which grossed $306 million in US theaters in 1996), *Twister* ($241 million in 1996), and *Deep Impact* ($140 million in 1998).[17] After a brief hiatus following the 9/11 attacks, the genre became even more popular. From 2004 to 2009, at least one disaster movie grossed over $150 million in the United States each year, and in all but one year there was a second feature that took in over $80 million.[18]

Box office figures do not fully represent the broad penetration of disaster themes into the public's consciousness, however. The majority of these films were summer blockbusters with large advertising campaigns that often featured central events of destruction so that even those who did not watch the films got their message. Many also became staples of cable television. Although industry figures are not available, the ubiquity of disaster films on cable television has been anecdotally noted by a number of commentators—for example, there is even a Facebook page called "I Always Stop If *Twister* or *Independence Day* Is on TV."[19] The number of witnesses who described the 9/11 attacks as reminiscent of a disaster movie also attested to their lasting impression. The film historian Geoff King noted that "some of those who told their stories" after the attack "likened the experience of trying to escape the dust and debris specifically to being inside a movie scenario," and Neal Gabler's editorial in the September 16, 2001, issue of the *New York Times* summarized the shared sense of deja vu: "The explosion and fireball, the crumbling buildings, the dazed and panicked victims, even the grim presidential address assuring action would be taken—all were familiar, as if they had been lifted from some Hollywood blockbuster."[20]

From 1990 to 2010, disaster films were increasingly related to the environment. Developments in computer-generated imagery (CGI) technology, growing evidence of anthropogenic climate change, and the end of historical antagonisms led filmmakers to cast natural phenomena as their villains. As Benjamin Svetkey put it: "We have no more Russians, no more Germans, no more villains. So we turn to mother nature."[21] In films such as *Twister* (1996), *Volcano* (1997), *The Day After Tomorrow* (2004), *Poseidon* (2006), *The Core* (2003), and *The Happening* (2008), nature was portrayed as a violent, destructive force fundamentally opposed to humankind. In *The Day After Tomorrow*, a fatal subzero frost chased Sam Hall (Jake Gyllenhaal) through the halls of the New York Public Library, while nature struck back at polluting, climate-changing humans in *The Happening* by releasing airborne neurotoxins that induced its victims to immediately (and quite creatively) commit suicide. Even when the disaster took the form of an alien invasion, many films still contained environmental elements. In some, aliens were seen as a part of the natural world—in *War of the Worlds* (2005), the invaders literally emerged from the Earth, where they had been buried for millennia. In others, aliens were representatives of an otherwise mute biosphere—in the 2008 remake of *The Day the Earth Stood Still*, Klaatu (Keanu Reeves) declared, "If [humans] die, the Earth survives," and nearly wiped out humanity in order to save the planet. While many Americans continued to deny the reality of anthropogenic climate change, increasingly dire scientific reports clearly dovetailed with the eco-apocalyptic themes of these films.[22]

Disaster Patriotism

Most critics are in agreement that there are broad social, political, and cultural currents responsible for the disaster genre's ebb and flow over time. As Stephen Keane put it: "Whether set in the past or extending to the future, disaster films carry the ideological signs of the times in which they are made."[23] For most film historians, the genre's rise to fame in the 1970s was closely connected with the widespread sense of national disillusionment during the Nixon-Ford era.[24] Especially in its most recent iteration, the epic scope of disaster films make them uniquely positioned to serve as bellwethers of the nation since major symbols of national pride, such as the Empire State Building and the White House, often play a symbolically crucial role in their stories. Since *Independence Day*, the wanton destruction of national landmarks has become a standard feature of the genre. "The implications," one critic noted, "are that the destruction of a landmark is enough to symbolize the end of a city and extending to the invasion/destruction of an entire country."[25] Government officials have often served as central characters in these films, and from 1990 to 2010 the president himself was increasingly cast in the ensemble.[26] In these ways and others, disaster movies invite an interpretation of their stories as national commentaries.

A number of critics have pointed to schadenfreude, the sadistic pleasure some find in witnessing pain (or scenes of destruction), as an explanation for the popularity of disaster narratives. Discussing *Volcano* (Mick Jackson, 1997), David Denby observed: "[The audience is] put in the position of wanting to see the people melt (otherwise there's no movie); we root for the disaster."[27] This is certainly a point of interest for some audience members, but I contend that the long-standing appeal of disaster movies is to be found not in the scenes of destruction themselves but in their characters' heroic responses to existential threats. Disaster films remain one of the most formulaic genres, and viewers can expect a prescribed sequence of events in which everyday life is suddenly disrupted by an unexpected event, centered in the United States, that will kill thousands, millions, or billions of people.[28] Characters from different backgrounds will be thrown together. They will not be able to prevent the catastrophe, but they will overcome conflicts, survive, and save others. Villains will perish, and some sympathetic characters will sacrifice themselves in the name of the greater good. Immature husbands will be transformed into heroes as crisis enables the reestablishment of proper (heteronormative) priorities. In contrast to action films, which also feature expensive CGI demonstrations of violence and destruction, disaster movies typically depict an ensemble cast of characters pulling together to affirm the importance of the collective. As Keane noted, "the innocent victims in disaster movies can only find strength in group action," which affirms the necessary bonds of the nation.[29]

In post–Cold War disaster movies, the ensemble cast is expected to be representative of America's diversity. Individuals of different ethnicities, races, classes, professions, and levels of education are forced to work together, often for the first time. In the face of crisis, the better angels of our natures prevail, and previously intransigent divisions dissolve. As such, race plays an especially significant role in this genre. Despite recent claims that the United States has entered a "postracial" era, racial divisions remain the most persistent fissure in an increasingly heterogeneous population.[30] Compared to more complicated intersections of categories of identity, the bridging of racial differences is relatively easy to show on film, through scenes of interracial collaboration and friendship. In the typical disaster movie, Americans prove themselves not only tolerant and inclusive but also genuinely colorblind. This symbolic unification is often not particularly subtle. In the conclusion to *Volcano*, for example, as gray volcanic ash is covering the citizens of Los Angeles, a city whose racial tensions and violence had been the subject of the national spotlight during and after the Rodney King verdict only a few years before, a small boy delivers one of the genre's overriding messages: "Everybody looks exactly the same."

The federal government's failure to aid black residents of New Orleans in the wake of Hurricane Katrina might have been expected to push directors to revise the disaster formula to reflect the unequal distribution of environmental risk according to race and class, but post-Katrina films such as *The Happening* portrayed a similarly unified, idealized response to environmental crisis. Imagined disasters, in this sense, serve an important national purpose; they provide, as Despina Kakoudaki pointed out, "a major incentive for the revival of humanist notions of community and patriotic identification" by creating "a common ground, a shared national point of view."[31] With the exception of a few notable films, most disaster movies end on a hopeful note of lessons learned, with an expectation of political, cultural, and spiritual rebirth as the credits roll. Although lives have been lost and national symbols destroyed, the cataclysm provides collective national regeneration through crisis.[32]

Reading the Disaster Movie: Peakist Narratives and Popular Culture

A genre as consistently popular as the disaster film succeeds because of its polyvalence. Some viewers may enjoy the use of CGI technology to create sublime spectacles, while others appreciate the unity, resolve, and heroism that most characters display in response to an imagined crisis. When scholars discuss the potential media influence of popular culture, they often portray audiences engaged in "decoding" in a "minority" (or "oppositional") way that is subversive and sometimes

not intended by the author, director, or artist.[33] However, an oppositional reading can also represent a literal interpretation of the material.
While *Jaws* (1975) has been seen as a film about submarines, patriarchy,
and the alliance between the forces of law and order and corporatism,
for the minority of viewers suffering from selachophobia its primary
lesson concerned shark attacks.[34] Similarly, while the dominant reading of disaster films may have little to do with alien attacks or falling
meteors, for viewers who were immersed in peak oil culture or deeply
concerned about climate change, the interpretation may be quite literal. While familiar categories of identity, such as race, class, gender,
religion, and politics, certainly influence the meaning that audiences
make of specific texts and genres, social or psychological factors—
beliefs about sharks, for example, or the likelihood of an imminent
ecological crisis—can be just as determinative.

Many peakists were avid fans of apocalyptic narratives across all
media types, including disaster films. Although a majority of Americans of a certain age have probably seen *Independence Day*, it is doubtful that nearly as many nonpeakists have seen less popular films such
as *The Postman* (which 40 percent of my American survey respondents
watched), *Waterworld* (65 percent), or *Children of Men* (36 percent) (see
table 1). Films of disaster and destruction became popular from 1995
to 2010 for some of the same reasons that the peak oil movement
coalesced, but they also seem to have affected the way that peakists
thought about themselves, their lives, and the future. Their readiness
to discuss the relationship between their beliefs and these fictional narratives presents a rare opportunity to investigate the impact of apocalyptic popular culture on real-world beliefs.[35]

When asked to comment on their interest in the disaster genre and

Table 1. Peakists and recent Hollywood disaster films

Film	Year	Box office gross (US theaters only, in millions of dollars)[a]	Percentage of peakists who have seen the film
Independence Day	1996	306.1	78.5
Mad Max	1979	8.4	75.2
An Inconvenient Truth	2006	24.1	69.8
Waterworld	1995	88.2	65.2
The Day After Tomorrow	2004	186.7	57.2
War of the Worlds	2005	234.3	53.8
The Postman	1997	17.6	38.6
Children of Men	2006	35.6	36.5
Cloverfield	2008	80.0	22.1

[a] Data provided by Box Office Mojo, http://www.boxofficemojo.com.

its potential connection to their belief in an imminent real-world col-lapse, peakists' responses reflected the wide range of ways that people understand their relationship to popular culture. Several granted that they enjoyed the genre but denied that it might have any impact on their beliefs, such as the Connecticut man who insisted that "the seri-ous peak oilers are serious people, not given to fantasy (as entertaining as that might be)."[36] Some confessed that they had actively resisted al-lowing pop cultural narratives to influence their view of the future but acknowledged that they had been "primed" to consider scenarios from disaster movies by years of exposure to these narratives.[37] An Ohio woman said, "I think it's hard not to be influenced by gloom and doom portrayals," and an oil company worker noted, "It's difficult to avoid letting fictional portrayals influence your thinking about situations with so many variables." A Washington man agreed that "it's hard not to be influenced by pop culture images" but did his best "to research the topic and develop a rational view of what is likely to happen."[38]

However, most respondents admitted that such attempts had prob-ably failed since, as a science fiction fan put it, "[human beings] are sponges, and cannot help but be influenced by media to which we are exposed." An Arizona man commented that these narratives were a "huge influence on the 'what to do about peak oil' issue," while a scien-tist in his fifties said that he had "read so many post-apocalyptic novels and seen even more movies that it would be difficult to deny that there has been an influence."[39] Indeed, studies of media effects regularly show that exposure to popular culture influences viewers despite their skepticism, media literacy, or cultural sophistication. For example, a team of British researchers found that, after watching *The Day After Tomorrow* in 2004, audiences were 50 percent more concerned about climate change than they were before the film.[40]

While countless scholars have discussed to the prevalence of apoca-lyptic narratives in contemporary American culture, the relationship between popular culture and end-of-the-world beliefs has rarely been documented directly. These survey responses suggest that cultural rep-resentations of apocalyptic themes can shape the direction of deeply held beliefs. This is "the end of the world as we know it" as a feedback loop, where narratives of the apocalypse—whether in film, literature, genre fiction, comic books, or video games—intermingle with and in-fluence the worldviews of actual people.

In the case of the film *Collapse* (Chris Smith, 2009), this media in-fluence came full circle. A documentary about the peakist Michael Ruppert, the author of *Crossing the Rubicon: The Decline of the American*

Empire at the End of the Age of Oil (2004), *Collapse* gave Ruppert ample screen time to explain the theory of peak oil even as it suggested that his obsession reflected an underlying emotional instability.[41] For some viewers, the film introduced a new idea that proved persuasive. On the blog Inspired Ground, for example, one peakist chose *Collapse* as the singular film that had most influenced her life.[42] Another, in a review on Amazon.com, described it as "a turning point in my life": "It has changed me forever. . . . If you read this Mike, I want to say: thank you for your legacy, for changing my life for the better and preparing our own little lifeboat for my family and I. I've been spreading your words ever since to whomever will listen."[43]

A number of survey respondents explained the media influence of disaster films by referring to the lack of alternate (nonapocalyptic) visions of environmental change, which assured that Hollywood films would have some measure of influence. A thirty-year-old man from California bemoaned the lack of competing, serious, and plausible mass-culture depictions of the future, asserting: "We need artists to describe for us a positive vision of the future." By *positive vision*, he meant not utopian dreams of technological salvation but narratives of a nation or planet dealing with urgent ecological problems in a serious, thoughtful, and concerted way—which, as of this writing, might sound as unlikely to some readers as the plots of *Armageddon* (1998) and *The Core* (2003). Similarly, a white male from Washington, DC, noting the connection between fictional and political narratives, argued that the absence of peaceful, nonviolent visions of the future was the cause of the peak oil movement's apocalypticism: "Unfortunately, because our leaders and our society have refused to conduct an open, honest, thorough discussion about this issue, we are left to our own wild imaginations and the imaginations of Hollywood marketers to develop images of likely futures."[44]

The Influence of Disaster Films on Peakists: Nature as an Agent of Change

How did disaster narratives shape beliefs about peak oil? For some respondents, disaster movies simply dramatized and confirmed existing suspicions—a woman in her sixties wrote that "disaster movies have reaffirmed [her] beliefs," while a man from Oregon claimed that "the portrayals vindicate deep-seated beliefs." A Californian said that such films "are all related to a common idea of what [the future] would be like in our imaginations" so that "art is imitating life." Another believer, who had attended a number of peak oil conferences, wrote: "These visions are persuasive in part because they tap into human imaginative habits that are common to many persons (that's why the movies get made)." Other respondents believed that disaster narratives had opened their minds to potential futures that just might come to pass. A Colorado man in his twenties noted that, while "fiction can be a cathartic way to release emotion & fear around outcomes," it "can also be prophetic," while a prepper from California reported that his familiarity with disaster films had influenced the way he conceives of the future: "I haven't been convinced of [sic] one in particular, but they do expand the realm of possibilities." Similarly, an Oklahoma woman in her forties said that "in some ways they have raised fears that might not have been there," while a Minnesotan in her early thirties summed up the general influence of apocalyptic popular culture: "I think it has made me more of a doomer."[45] Indeed, American peakists, who had watched many more Hollywood disaster films, were noticeably more extreme in their predictions than believers in other countries. They were much more likely to see resource wars and an apocalyptic scenario (including a global die-off and civilizational collapse) as probable future events than were their Canadian or European counterparts, for example.[46]

In most disaster films, nature is portrayed as a partial agent of social and political change. A tornado, a tidal wave, or an alien attack is the prime mover of each narrative, compelling humans to unite across political, racial, and socioeconomic boundaries. For peakists, naturally occurring scarcity plays a crucial role in the kind of positive transformation evident in many peakist narratives. Although many believers saw the consequences of peak oil as a slow, "long emergency," some spoke of the peaking of global petroleum production as an event with destructive capabilities, akin to a tornado or an earthquake. For example, one respondent wrote that "peak oil will have a permanent and measurable effect on [his] day to day living," and another maintained that "peak oil will drag down any economic systems that remain dependent on fossil fuels."[47] Whereas most environmentalists now see resource scarcity as tightly bound to economic and social issues that are highly variable, peakists tended to hold fast to a version of the limits-to-growth environmental paradigm discussed in chapter 2, where economic and social issues are at the mercy of ecological limits. The relationship between the natural environment and human civilizations in both disaster and peak oil narratives can be summarized by an old adage: "Nature bats last."

In both disaster films and peak oil narratives, hope is to be found in disaster's wake since cataclysmic events provide a means of regeneration through crisis. In disaster films, tectonic plates, volcanoes, and even flora are potential villains—primal powers that threaten to destroy lives and nations alike. In response, humans are forced to pull together, and, ultimately, this unification is more consequential than the material damage and lost lives. Disaster forces survivors to recognize their true priorities in life and, thereby, provides a means of regeneration through crisis—under attack from nature, old friends become lovers, workaholics become dedicated parents, and families unite. Even darker films such as *The Core* (2003) and *The Knowing* (2009) conclude not by focusing on death and destruction but by highlighting the lessons learned and the strength of the survivors as the soundtrack swells.

What, then, is the peakist vision of the future? On peak oil Internet forums there is a great deal of speculation, prediction, and prophecy about the future—the "Planning for the Future" section of one Web site alone contains nearly 3,000 separate threads, some of which have been viewed over 300,000 times.[48] As Daniel Wojcik observed, "the privatization of apocalyptic beliefs" in recent decades means that "the consumers of [mass-marketed items of apocalyptic prophecy] may never come into direct contact with one another but instead may construct a personal vision of apocalypse gleaned from a diversity of sources."[49] By *privatization*, we should understand *individualization*—as is evident from speculation on peak oil Web sites, apocalyptic beliefs, especially in the Internet age, are highly personal and often idiosyncratic. This variance can make it difficult to speak with any accuracy of a general "peak oil vision." However, as peakism grew in popularity, a number of adherents wrote speculative novels about the future that proved influential within the subculture. Nestled within the literary genre of science fiction, peak oil fiction sheds light on peakists' imagination and actions.

Petrofiction and the Great American Oil Novel

Amitav Ghosh noted in 1992 that, despite the ubiquity and importance of oil in the contemporary world, petroleum was conspicuously absent in fiction. In claiming that "the Oil Encounter . . . has produced scarcely a single work of note," Ghosh was referring to the transnational exchange between Middle Eastern producers and Western consumers, but he pointed out that even within the West there is a noteworthy absence

of "petrofiction," let alone a "Great American Oil Novel."[50] Two decades
(and over 500 billion barrels of oil consumed globally) after that proclamation, the genre of petrofiction emerged from the American peak
oil community.[51] Novels such as *After the Crash: An Essay-Novel of the
Post-Hydrocarbon Age* (2005) and *Oil Dusk: A Peak Oil Story* (2009), which
place petroleum scarcity at the center of their secular apocalypse and
post-peak worlds, speak directly to Ghosh's lacuna. By demonstrating
the collapse that results from the cessation of the transnational flow of
oil, they provide an implicit critique of American petrodependence. As
of this writing, more than a dozen peak oil novels have been published
in English.[52] They were written by dedicated peakists, marketed as peak
oil novels, and recognized as such on peak oil forums.[53]

As do disaster films, novels such as *Last Light* (2007), *Boil Over: The
Day the Oil Ran Out* (2010), and *Shut Down: A Story of Economic Collapse and Hope* (2011) focus on an ensemble of diverse characters as
they respond to the unexpected—but predictable!—collapse of the
world around them. In peak oil fiction, however, narrative takes a back
seat to ideology. The narrator intermittently breaks from the plot to
educate the reader about the process by which energy depletion precipitated the disruption of the status quo, a scramble for survival, and,
ultimately, the fall of civilization. Given their similarities, these novels
are best categorized by where they locate themselves on this fixed time
line. Although most are written after the crash, a few portray the period just before it and document the transformative process of becoming peak oil aware.

Prelude (2010) is one of these. Written by Kurt Cobb, a frequent con-
tributor to The Oil Drum and Energy Bulletin and a founding member
of the American chapter of the Association for the Study of Peak Oil
(ASPO), it was dedicated "to the peak oil community" and contains a
number of scenes that peakists would find familiar, such as local meetings of energy study groups and a national conference (presumably
that of ASPO). Its primary narrative is the education and conversion
of its protagonist, Cassie, a naive upstart at an energy consulting firm,
from a "cornucopian" to a true believer. As if to answer Ghosh's challenge, peak oil novels attempt to reorient the reader's understanding of
the world so that all that is solid melts into oil:

The products in the store windows she was passing had arrived by truck, train and
ship. The camera shop, the electronics place, the women's clothing store, the phar-
macy, and the video store she passed, none of them would be operating without oil.
Many of the products and even their packaging were made using materials based

on petrochemicals, materials like plastic and synthetic rubber. Even the perfume that was emanating from the woman walking in front of her was probably partly derived from oil. And the cars and trucks and buses coursing through the street beside her all ran on some form of oil. . . . Suddenly for Cassie the whole world had now become one big manifestation of energy, much of it in the form of oil.

By highlighting the hidden petroleum that lubricates and enables our modern lives, *Prelude* achieves a cosmological reorientation that places energy at the center of contemporary life. Cassie notes, "This was the world that oil built," and wonders, "How long will it last?"[54] Her personal struggle is intended to spark and mirror a similar process of discovery and education in the reader. As if in preparation, the novel describes the sense of conversion, emotional turmoil, and acceptance that we witnessed in chapter 1.

Novels such as *Boil Over* (2010), *Deadly Freedom* (2008), and *Oil Dusk* (2009) occur in the midst of the collapse and, like disaster films, expend little effort on character development. *Last Light* (2007), published by the British author Alex Scarrow, is the most well-known of these works.[55] Scarrow's Sunderland family experiences a familiar series of events: the collapse of the state, the formation of militias, and the eventual isolating localization. *Last Light* ends on an uncharacteristically optimistic note: in a scene that echoes the denouement of the 2005 remake of *War of the Worlds*, the Sutherland family reunites in their suburban home with a glimmer of hope that although "the oil age is over . . . things will eventually knit themselves back together again."[56] Andy Sutherland, the family patriarch, is their savior—resourceful, clever, and strong, with military training as an engineer in Iraq that serves the family well.

Andy Sutherland is also, significantly, a peak oil prophet. Most peakist novels (and disaster films) contain a Cassandra whose warnings go unheeded; typically a white male (like most peakists), this character is better prepared than others when disaster strikes and, thereby, survives along with his family. This is a key aspect of many apocalyptic texts that has been ignored by most scholars—the prophet serves as a model for the reader, who can now imagine himself as more prepared for these events.[57] In an author's note, Scarrow makes this function explicit: "I'd like to think that a whiff of *Last Light* will remain with you once you snap the cover shut. I'm hoping Andy Sutherland achieved something; that the world looks slightly different to you now—more fragile, more vulnerable. After all, *to be aware is to be better prepared*."[58] Indeed, a peakist from Kansas pointed to this link between fiction and psychological disaster preparation when he wrote that "some of these portrayals" in apocalyptic popular culture "can appear to be almost heroic in nature": "It would seem some [peakists] think they could be the 'last one standing' because of their prep[aration]s, but I think that is natural at some level." And a Texan acknowledged why he found apocalyptic texts to be useful: "[They] let me play out scenarios in my head so that if/when confronted with similar situations, I can resist a mindless fight/flight//freeze condition."[59]

Scarrow's direct identification with his protagonist constructs a connection between prophecy and fulfillment that reflected a desire for crisis among many peakists. As participants grew more identified with the peak oil theory—discussing it with friends, family, and coworkers, speaking with other peakists on the Internet, and immersing themselves in subcultural books and films—the expected crisis became a longed-for event.[60] At least one believer noticed this tendency, writing on Peak Oil News: "Most of the ardent followers seem to really hope it will happen, either cause they despise the political/economic status quo or they've put so much time and energy into the issue that they feel the need to be proven right."[61] If it occurs, they are prophets and potential saviors; if it does not, they are merely "kooks" and "crackpots" who fell for a "fringe Internet-based theory."[62]

After Oil, a World Made by Hand

For the best exposition of the peakist ideology and reflection of the way that peak oil narratives differ from secular apocalyptic representations in popular culture, we should turn to the major peak oil novel, James Howard Kunstler's *World Made by Hand* (2008). As a result of Kunstler's notoriety, it was widely reviewed—in the *New Yorker*, the *New York Times*, and Slate, for example—and its author was catapulted to a new echelon of celebrity. Kunstler was even interviewed on Comedy Central's *The Colbert Report*, where he gamely explained to the host's conservative alter ego why "we're not going to run Walt Disney World, Walmart, and the interstate highway system on any combination of solar, wind, nuclear, ethanol, biodiesel, or French fried potato oil."[63] Along with his *The Long Emergency* (2005) and Richard Heinberg's *The Party's Over* (2003), *World Made by Hand* is one of the most influential peak oil works, read by 78 percent of my respondents. They found the novel to be not only "seriously addictive" but also prophetic—in their comments about it, many used the word *real* or *realistic*. A peakist from

Pennsylvania, for example, argued: "*World Made by Hand* gives quite a probable glimpse of what the post Peak Oil world could be like later this century."[64]

World Made by Hand is set in the mid-2020s, a decade and a half after the long emergency caused by peak oil: gas shortages that led to resource wars, social collapse, starvation, and lethal epidemics. Its primary location is Union Grove, a small town in upstate New York whose quaint downtown is now inhabited by a few score of aging townspeople living simply without petroleum and electricity. Its protagonist and narrator is Robert Earle, a middle-aged, white, liberal male, the "former CEO of a technology company" (and former peak oil prophet) who now works as a carpenter. Outside the town are three alternative living arrangements: Stephen Bullock's farm, a plantation; Karptown, where a band of neoprimitive roughnecks lives in shanties and trailers; and a religious cult, the New Faith Congregation.[65] The narrative center of *World* is a trip that Earle and a group of New Faithers make (by horseback and foot) to nearby Albany, which is controlled by a vicious strongman. The trip brings into relief the benefits of Union Grove, which has not suffered from war, nuclear attacks, or political corruption. The reader might be expected to use his foreknowledge to end up in Union Grove instead of Bullock's Farm, Karptown, Albany, or a mass grave.

While the novel has a number of similarities to disaster films, its differences underscore the ways that the peak oil movement departs from standard American apocalyptic narratives. I previously highlighted the symbolic importance of race in many disaster movies, and Kunstler also uses race as a symbol of the nation's direction. In *World Made by Hand*, a ragtag band of multicultural Americans does not come together to reaffirm the country's greatness in the face of disaster. While there are no people of color in the novel, Earle and others repeatedly refer to "race wars" in other locations. When the New Faith leader is asked why the congregation abandoned its previous location, he responds: "The white against black and so forth was spilling over from Philly, too, and we had trouble with it. . . . [T]hings are rough from sea to shining sea. From Texas clear to Florida, there's folks shooting each other and trouble between the races and all like that." Later, another New Faither pinpoints the reason: "Race trouble, to be honest. . . . A lot of people cut loose when Washington got hit [by a terrorist attack], you know. They left there with nothing but the clothes on their backs and some firearms. You had civil disorders in Philadelphia and Baltimore, refugees fleeing, what you folks call pickers, bandit gangs. Pennsylvania became a desperate place. After a while, it was like cowboys and Indians."[66] In this narrative, a terrorist attack does not unite the nation, as typically occurs in disaster movies, but instead brings long-standing divisions to the fore. The comparison of future racial violence as akin to "cowboys and Indians," a children's game dramatizing genocide, is a particularly pessimistic comment about the future of the Republic.

This seemingly reactionary perspective on the persistence of racial constructs and the explosion of racial violence in a dystopian future should be situated within the context of the novel's readership. As Democrats, leftists, and progressives, most peakists hold views on race that are markedly liberal. For example, in response to the question, "In recent years, has too much been made of the problems facing black people?" 10 percent of peakists answered "too much," while over a third answered "too little." In 2010 pollsters asked the same question of a representative sample of all Americans, and 28 percent answered "too much," while only 16 percent answered "too little." While peakists' prophecies may bear some similarities to racialized survivalist fantasies penned by conservative authors (such as William Luther Pierce), they should be understood differently.[67] Rather than a white supremacist fantasy, the post-peak world portrayed in *World Made by Hand* is a referendum on the nation's failure to achieve its founding promise of equality, which Kunstler suggests by juxtaposing "trouble between the races" with "from sea to shining sea" in the passage quoted above.

Almost all peak oil novels focus on the United States. For decades, Americans have conceived of oil as a distinctively American commodity—even when located below other countries—because of its role in facilitating the American way of life from the post–World War II period to the present. The geographer Matthew T. Huber has argued that "this entitlement is negotiated through a kind of livelihood discourse—or mobilizing a cultural claim to resources as crucial in sustaining a morally righteous way of life." The deep connection between oil and American identity is represented by and reaffirmed through the regular consumption of gasoline in stations with names such as "Liberty," "Freedom," and "SuperAmerica," often festooned with national flags, "a concrete, everyday practice through which Americans imagine their own national community."[68] Gas stations are part and parcel of the American way of life, which is characterized by movement, open roads, and the freedom of physical and social mobility.

The sedentary post-peak world, Kunstler suggests, is not only post-American; it is distinctly un-American. This is most evident in the setting of the novel, claustrophobic in its localism. Disaster movies, as well as classic disaster novels such as *Lucifer's Hammer* (1977), convey their national or global scope by incorporating scenes from as many different locations as possible, but *World Made by Hand* is, like its characters, unable to escape upstate New York.[69] Without petroleum for transportation, let alone electricity for the Internet, Earle and other survivors experience the paralysis of geographic isolation. Information about the US government—the identity of the president, news of the nation, and whether it even exists—is conveyed through the radio, which requires electricity. This is the nation as a truly "imagined community," and, when Earle mentions at the end of the novel that "the electricity stayed off, without even a few more additional spasms," we are left to assume that the United States has completely collapsed.[70]

Post-Apocalyptic Romanticism

Is this a crisis, or an opportunity? Although most reviewers described *World Made by Hand* as dystopic, its post-peak world is actually represented as superior to our present in many ways. This is no anomaly— one of three survey respondents admitted to feeling "excitement" about the post-peak world, while an independent survey of peak oil believers found that 37 percent were more "excited" than "worried" about the "post-peak years" and that 50 percent saw peak oil as an "opportunity," not a "problem."[71] In *World Made by Hand*, Union Grove's survivors have forged a genuine community founded on mutual aid. The crash of the oil economy has finally caused the fall of capitalism itself, and residents tend their gardens and employ a precapitalist credit economy, simple but fair. The government's absence has forced most survivors to embrace a Jeffersonian self-reliance, and they have gained the dignity of working freely for themselves. By necessity they have become renaissance people, producing not only their own food and housing but also their own alcohol, drugs, and entertainment—Earle is in Union Grove's bluegrass band, which covers golden oldies such as Nirvana's "Smells Like Teen Spirit." The survivors, as one character puts it, "eat real food instead of processed crap full of chemicals": "We're not jacked up on coffee and television and sexy advertising all the time. No more anxiety about credit card bills."[72] To emphasize that post-apocalyptic cuisine is actually superior to our own—no fast food, no chemicals, no trans fats—the narrator lingers on every culinary description. One reviewer observed: "You can practically taste the corn bread and the fish that choke this world's now-unpolluted rivers and streams, another upside to that whole end-of-civilization thing."[73]

Post-apocalyptic fantasies often engage in what Leo Marx called the *American pastoral ideal*, the "urge to idealize a simple, rural environment."[74] Henry Nash Smith, writing in 1950, noted the powerful influence that the dream of an "agrarian utopia of hardy and virtuous yeomen" had exerted throughout American history. He summarized the ideology inherent in agrarian utopianism in ways that are still relevant to contemporary post-apocalyptic fiction: "That agriculture is the only source of real wealth; that every man has a natural right to land; that constant contact with nature in the course of his labors make him virtuous and happy, . . . [and] that government should be dedicated to the interests of the freehold farmer."[75] With *World Made by Hand*, Kunstler situates his post-American novel within an American tradition. Robert Earle tells us, for example, that "the freehold farmer"—a central figure in a mythologized version of American history—"was the new chief executive." Kunstler's knowing nod to pastoral idealism is a complicated gesture, invoking the American mythos while prophesying the decline of the nation. If disaster movies and peak oil visions both provide a means of environmental and social regeneration through crisis, they part ways on the role of the nation. In the former, the nation plays a central reconstructive role in the post-apocalyptic world; in the latter, the US government disappears, but its absence does not seem to trouble anyone. Borrowing a motif from disaster films, one character mentions that "New York City is finished," a post-apocalyptic wasteland where "the immense over-burden of skyscrapers in Manhattan had . . . proved unusable without reliable electric service."[76]

Just as the pastoral ideal has served as a motivation for generations of environmentalists, many peakists believe that the post-peak world will be environmentally superior.[77] In Kunstler's vision, post-carbon survivors are able to enjoy a renewed sensitivity to and connection with the natural world. One character notes with pride that they "follow the natural cycles," and Earle praises the quiet life: "I enjoyed the peacefulness and easy pace of the walk. In a car, I remembered, you generally noticed only what was in your head or on the radio, while the landscape seemed dead." Kunstler deploys the pastoral ideal in his lyrical descriptions of natural landscapes. When Earle comes upon Britney, one of his lovers, gutting a fish:

She reached in and removed its guts and flung the guts out in the current. Then she ran her thumb down along the spine inside of the rib cavity to get out the congealed blood there that can make the meat taste off if you leave it in, especially on a hot day. Finally, she slipped the fish inside the creel and washed the spine and blood off her fingers in the current. I clapped my hands in appreciation. Hearing that, she finally turned around. What a sight she was in a wet cotton dress. I kicked off my boots and waded out in the water, scooped her into my arms and carried her back to the gravel bank.[78]

As this cinematic passage shows, the aftermath of the peak oil crisis gives Earle the opportunity to perform a classically masculine pose. Here and elsewhere, Kunstler adopts a century-old trope about a crisis of American masculinity caused by stultifying modernization, a topic that I explore in the next chapter. A return to a pioneer lifestyle provides a welcome change for the former CEO—and might for you too, it is implied, if you only read the signs and prepare for what is to come.[79]

Post-Peak Anticapitalism

A few scholars have analyzed *World Made by Hand*, but they have treated it primarily as work of fiction.[80] As a novel, it is a curiosity; as a work of prophecy that tens (if not hundreds) of thousands of Americans subscribed to, it takes on a greater significance. In the context of real-world believers, this vision of the future is most revealing when we consider the primary trait that sets most peakists apart from an average American: their politics. The peak oil movement has been characterized by journalists as the "liberal apocalypse" for good reason: 56 percent of American peakists described themselves as "liberal" or "very liberal," compared to only 21 percent of all Americans in a national poll conducted at the same time. Given a wider range of categories, approximately 7–9 percent would have considered themselves "anarchist," 10–11 percent "socialist," 20 percent "progressive," 14–15 percent "liberal," and only 4–6 percent "conservative."[81] Beyond labels, their values put them far to the left of most Americans. Eighty-one percent thought that gays and lesbians should be able to legally marry (with 11 percent having no opinion), for example, and 82 percent agreed with the statement, "We don't give everyone an equal chance in this country."[82]

In interviews, surveys, and online forums, many peakists advanced a leftist critique of American imperialism as well as of capitalism. They viewed patriotic American platitudes as a mix of misinformation and outright falsehoods. Judging from representative comments on virtual forums, many believed that "the American Dream is just propaganda used to justify selfish materialism" and that "the sad reality of the matter is that 'the American Dream' is dying": "Every month more American families are slipping out of the middle class and into poverty."[83] One peakist scoffed, in a thread on the subject of "American exceptionalism": "We have been the best colonialists in the world. We sucked the world dry, taking a full quarter of its resources for ourselves even though we only made up 5% of the population. And only a smidgen of that 5% actually got any of it."[84] Distrust of the US government extended to foreign policy, which most peakists considered to be motivated not by noble intentions but by short-sighted energy imperialism.

The number of survey respondents who described the United States as an empire was also telling. A few representative commenters considered oil depletion to be part of "the impending fall of the U.S. empire," caused by "Western Civilization's hubristic blindness and imperialistic self-destructive tendency," which would lead to "the end of US imperialism." Even more cited the imminent peak oil collapse as the death knell of capitalism. A New York man in his fifties noted that "capitalism requires constant *growth* [sic] for it to remain viable as an economic system, but that continuous growth may well no longer be possible due to peak oil," while a peakist in her early forties observed that "capitalism as we know it has grown on the back of cheap energy, particularly oil."[85] As a commenter on Peak Oil News put it, "five dollar gasoline is great for mother earth" not because it will stimulate the development of alternative energy sources but because "it means the end of capitalism."[86] These predictions are represented in peak oil fiction, where most authors foresee local systems of mutual exchange replacing neoliberal capitalism. As the cultural theorist Imre Szeman asked: "Is the end of oil a disaster? This depends, of course, on the perspective one has on the system in danger of collapse: capitalism."[87]

Peakists' politics should alter our evaluation of their vision of the future. While they do not present the radical racial reconstruction of an Octavia Butler or the nuanced gender critique of an Ursula K. LeGuin, these Americans are engaged in their own radical cultural project of imagining a post-American, postcapitalist space.[88] It would be an overstatement to paint all believers as conscious anti-imperialists or anti-capitalists, but the close connection that most draw between imperialism, capitalism, and petroleum shows that they view the consequences of peak oil as America's proverbial chickens coming home to roost. One article by a reformed peakist claimed, with some exaggeration, that "loathing for the United States is a virtual prerequisite for becoming a peak oil acolyte."[89] Like many others, the Virginia man who predicted that "our American way of life is coming to an end" seemed to view the end of the oil age as the end of the American empire and capitalism itself: a crisis, of course, but also an opportunity for something better.

From Politics to Prophecy: Political Quiescence among Peak Oil Believers

While beliefs about inevitable futures have often inspired action, they can also lead to political quiescence. Instead of becoming more engaged in radical or environmental politics, the average peak oil believer became less politically active. Although they followed national and environmental news more closely after becoming peak oil aware, peakists became less likely to vote or attend marches, rallies, or protests.[90] What is most striking about this group of leftists, then, is their tendency to see even radical politics as unable to solve environmental problems. Anxieties about energy motivated a number of political campaigns in recent American history, such as the antinuclear movement of the 1970s and 1980s, but peakists almost completely ignored electoral politics as an avenue to address their concerns. One representative respondent described himself as "cynical and disassociated": "Our government is unresponsive, in denial, and ineffectual." A Colorado man asserted that "politics is a supreme waste of time," and an Oregon man complained that "our political system and institutions are dysfunctional to the point of being irrelevant—the emperor has no clothes."[91]

For many people, their new ecological identity occasioned a retreat from politics. A Texas man "briefly got into politics," voting in primaries and attending county caucuses and conventions, but then "realized that the system didn't want to change" and "was not likely to change" without "some external crisis."[92] Oily Cassandra, whom we met in the introduction, had previously been active in leftist politics, such as protests against the US invasion of Iraq in 2003, but reported in 2007 that there was no place for her concerns in mainstream American politics.[93] James Howard Kunstler regularly critiqued American politics on his blog, but he did so from an outsider's perspective, claiming that political action was ineffective given the "self-righteous cluelessness in every band of the American political spectrum."[94] Kathy McMahon (the Peak Shrink) had been active in the civil rights movement, but Peak Oil Blues did not advocate political action. She viewed the problems of petroleum dependency as far too ingrained for contemporary politicians to address and informed her readers: "[It is] not your job to fight [the Panglossian disorder]. You cannot. . . . Maybe this year, maybe next, you may be thought of as the sane one after all."[95]

As these comments suggest, peakists were even more skeptical than average Americans were of elected politicians—only 3 percent of respondents said that most members of Congress deserve reelection, compared to 21 percent of Americans at the time of the survey.[96] This skepticism is reflective, in many ways, of the "sense of alienation" consistent with a default libertarianism, which a recent CBS poll found "had ticked up in recent years": "Most Americans feel alienated from their political leaders and dissatisfied or angry with Washington."[97] When the historian Timothy P. Weber pointed out that "advocates of apocalypticism" have typically been "outsiders, alienated and disinherited from the privileged and powerful," who "looked for their future redemption from beyond the clouds precisely because they had no recourse in the present," he was discussing how marginalized groups embrace extreme religious ideas, but we might very well apply his insight to our (more) secular subject.[98] A core aspect of American political culture in the twenty-first century is a general feeling of political alienation—in a 2011 poll, 73 percent of respondents agreed with the claim, "The people running the country don't really care what happens to you," and 66 percent agreed that "what you think doesn't count very much anymore."[99] In a different poll, Americans placed more faith in telemarketers than in elected congressional representatives.[100] This sense of alienation has long been reflected in the low US voter turnout compared to that in other industrialized democracies.[101]

Peakists' surprising alienation—their belief in the futility of politics—was fueled by an awareness of the need for radical changes in environmental policy combined with a relatively sober calculation of the limitations of contemporary American environmental politics. As the United States underwent its sharp turn to the right over the last three decades (as described in the last chapter), leftist calls for more government intervention (beyond health care) have become politically marginalized. Even with the emergence of the Occupy movement in 2011 and the reelection of a Democratic president, most leftist political goals remain beyond the pale of mainstream American political culture in the Age of Reagan.

This is especially true of environmental initiatives. Although they overestimated the timing and consequences of conventional petroleum depletion, peakists' sense of an imminent ecological crisis is certainly justified—one could point, as they do, to the grave threats posed by climate change, ocean acidification, topsoil erosion, deforestation, and environmental toxification. Their evaluation of the state of American environmental politics, while perhaps overly defeatist, is not out of touch with the candid assessments of many mainstream scientists and environmentalists. Most peakists argued that the United States is simply unable to solve the major problems that the country and the planet face, problems that they identified as peak oil, climate change, and capitalism's demand for infinite growth in a finite world. This belief was summarized by a Nebraska man in his thirties who believed that political action on these fronts was futile: "[The] social, political, and media systems required to make meaningful headway are dysfunctional. Appropriate responses require a level of discussion that is beyond the capacities of the present [US] system."[102] As these Americans "peer into the uncertain future," as one member of Peak Oil News put it, "they find it easier to believe in the complete disintegration of America and its culture than in the possibility of an American society which has adapted to changed circumstances and innovated new solutions."[103] Given the preponderance of apocalyptic narratives in American culture and the absence of alternate visions of environmental and social change, this perspective should come as little surprise.

Since fossil fuels are finite resources and the United States is unlikely to suddenly transition away from its petroleum economy, might not oil depletion, the thinking goes, deliver a solution to (or salvation from) climate change and ecological crisis? If this seems like mere wishful or magical thinking, we might ask ourselves which scenario seems more likely: that American politicians, including the House of Representatives, along with leaders from other populous and high-energy-consuming countries, will collectively decide to drastically reduce their energy usage and convince or require oil companies and petrostates to forgo burning every barrel of petroleum, at great economic loss, or that they will be forced to do so only when they finally run out of cheap oil?[104] This line of argumentation is many things—pessimistic, defeatist, disheartening—but it was neither uninformed nor unrealistic. In fact, it resembles the claims made by advocates of climate adaptation and geoengineering.[105]

For those who hope that the United States will take a more active role in addressing pressing energy and environmental issues such as our dependence on fossil fuels, the fatalism and political quiescence of peakists is troubling. As a group of educated, middle-class leftists who based their very identities on the threat of resource depletion and environmental destruction, peak oil believers might be expected to be at the forefront of demands for changes in environmental policy. Instead, they became more fatalistic, less politically active, and more quiescent. They noted, with a touch of black humor, that peak oil might provide a solution to anthropogenic climate change. In some ways, this was a self-fulfilling prophecy: if the most environmentally aware and concerned citizens retreat from the public sphere into individualistic prepping or even into collective groups such as Transition Towns, the public pressure for political action will only diminish.[106]

Many scholars have viewed narratives of ecological apocalypse in opposition to radical politics, but the relationship is more complicated.[107] Claims that eco-apocalyptic attitudes lead to political passivity may be true, but we might also view the peak oil movement as the sublimation of a political vision into a prophecy or a new configuration of radical political beliefs that reflects our privatized, neoliberal age. From this perspective, we can better understand the phenomenon's concentration in the United States, where most government bodies have been much less willing to even acknowledge the reality of one ecological crisis (anthropogenic climate change) and pay more than lip service to the country's reliance on petroleum. Peakists' fatalism is disquieting not because it is puzzling but because it is so recognizable—from their political alienation, to the influence of apocalyptic popular culture, to their suspicions about the futility of politics. Scholars such as Constance Penley have noted that the popularity of science fiction (and speculative fiction such as *World Made by Hand*) is directly related to our ability to imagine meaningful social change. She argued that popular narratives often embrace messiahs or disasters as agents of change because "we can imagine the future, but we cannot conceive of the kind of collective political strategies necessary to change or ensure that future."[108] Most peakists—and perhaps even most Americans—would agree.

FIVE

White Masculinity and Post-Apocalyptic Retrosexuality

Compared to my great-grandmother, I marvel at the life I have had. . . . I have roamed North America at will from Ixtapa to Halifax. I eat like a king. I want for nothing. I know that my ancestors were more connected to nature, that they were survivors, but I am wealthy and soft. WHITE MALE, OREGON, THIRTY-SIX TO FORTY

This isn't like sharing chores. It's a matter of life and death. Political correctness goes out the window and he (or she) who demonstrates better competence as a warrior gets dibs. What that means is Ripley [from the film *Alien*] comes before [Steve] [U]rkel [the iconic nerd from the television show *Family Matters*], and Rambo [from the *Rambo* films] comes before Paris Hilton. "THE RETURN OF PATRIARCHY," PEAK OIL NEWS AND MESSAGE BOARDS, FEBRUARY 14, 2010

On the Web site Transition Culture, one member asked, shrewdly: "What if 'peak oil' isn't really about the powerlessness people fear in any near future, but is actually about the powerlessness they feel today?"[1] The topic at hand, however, was not contemporary American political quiescence but gender in the peak oil movement. Indeed, in statements like those reproduced as this chapter's epigraphs, the excitement that some peakists felt about the post-apocalyptic future certainly seemed to be tied to gender and race. In my surveys, between three-quarters and four-fifths of all peakists were white men, and close read-

129

ers will have noticed that nearly every peakist author I have referred to is a white male. Indeed, one well-known author reported: "The attempt to find a female speaker at a peak-oil conference is a disaster. I tried hard with ASPO-5 [in 2011], but no luck."[2] Why were such a high percentage of peak oil believers white men?

Peakists themselves were interested in and sometimes concerned by this gender imbalance. In the comments section of the same Transition Culture blog post, they offered a number of potential explanations. A mother in her thirties cited the gendered division of labor as one factor, noting: "Time is . . . a very real issue for every woman I know. Working, taking care of children, trying to build community, attending to all the details of daily life—this is exhausting. Women do a huge amount of necessary work in this world." "ChristineL" agreed, citing the "old feminist slogan" that goes, "I wanted to go out and change the world, but I couldn't find a babysitter." "Greg" mentioned differential gender "socialization," arguing that "women are generally awarded for being team players" and "following the herd" but that communicating about or preparing for peak oil "asks that we go against the flow" of cultural norms "and do the hard thing": "Men are generally more willing to do that." Matt Savinar, then the Web master of the popular site Life After the Oil Crash, added the sociobiological angle that status seeking may have something to do with men's willingness to speak out: "Men are a bit more outspoken and like to post all the great stuff they are doing on web blogs to get attention. . . . Tribesman [sic] in New Guinea have their penis sheaths. We have our Peak Oil blogs." Less speculatively, in a 2007 blog post entitled "Is Peak Oil a Guy Thing?" Kurt Cobb concluded that the dearth of female peakists is the consequence of "the peak oil movement draw[ing] many of its members from the oil industry," which is "a highly technical subject which attracts minds from the hard sciences, engineering, mathematics, and the high technology world, all of which continue to be dominated by males."[3]

While gender socialization, sociobiological tendencies, and the history of employment discrimination may have contributed to the discrepancy in participation in the peak oil movement, many collective visions of post-apocalyptic life, as depicted in peakist fiction and discussions on virtual forums, suggest that the real answer lies at the nexus of culture, political economy, and psychology and has much more to do with men than with women. In this chapter I argue that, for some white male participants, the peak oil movement and similar survivalist phenomena provided one means of reinscribing "traditional" (read: 1950s) gender roles and revitalizing white masculinity. After exploring the depictions of and tensions around masculinist post-peak fantasies within the peakist community, I explore their congruence with popular survivalist fantasies in twenty-first-century American culture, such as zombie attack scenarios, via the television show *Revolution* (2012–14). I situate these anxieties over white masculinity within the broader context of transitions in political economy, gender, and racial scapegoating in the twenty-first century.

As noted in the last chapter, women are minor players in and people of color are entirely absent from *World Made by Hand*, but the revitalization of white masculinity is a central concern. Robert Earle, the novel's narrator and protagonist, is a former white-collar worker who finds a renaissance masculinity in the post-apocalypse. Throughout the course of the novel he inhabits a number of models of twenty-first-century masculinity. When he toils as a carpenter, he is a blue-collar worker; when he makes a trip to Albany and kills a man in a daring escape, he is a soldier and an action hero; when he saves a young widow and a child from a fire, he is a fireman; when he is voted mayor and sheriff, he is a lawman; and, when the town celebrates, he is a musician, if not quite a rock star. In his late fifties or early sixties, Earle also enjoys a striking amount of sexual freedom. As the novel begins, he spends his nights with Jane Ann, whose husband, Loren, is impotent. After a young man in town perishes in a fire, Earle reluctantly accepts the attractive young widow Britney into his home. This is described as a selfless act of generosity—post-peak Union Grove is simply no place for a single woman, of course.

Female characters do not enjoy the same opportunities, and the gender politics of *World Made by Hand* provoked an uproar within the peakist community. None of its central characters are women, and it would flunk the famous "Bechdel test" that identifies gender biases in works of fiction.[4] The two primary female characters (Jane Ann and Britney) seem to exist, as the peak oil feminist Sharon Astyk put it, to "a. compete to have sex with the narrator and b. suffer and c. serve meals."[5] Kunstler's narrator justifies these regressive gender roles matter-of-factly: "As the world changed, we reverted to social divisions that were thought to be long obsolete. The egalitarian pretenses of the high-octane decades had dissolved and nobody even debated it anymore, including the women of our town."[6] A number of female peakists objected to this prediction, noting that the gains of feminism were not likely to be undone within a decade or two: "I very much doubt women will lose political representation and be forced into awkward, bulky dresses a mere two decades or so after collapse."[7] The blogger

Janet McAdams wrote: "What creeped me out about the book was that [Kunstler] seemed to romanticize the vulnerable status of women in his post-[peak] world as part of a project of re-heroizing 'good men.' It felt paternal, vaguely Humbertish."[8] Another put it more plainly: "For god's sake can't we just come right out and say it? Kunstler is sexist pure and simple . . . [and his novel is] an utterly ridiculous demonstration of old-white-guy fantasies gone wild."[9]

Forecasts of a post-apocalypse that is also postfeminist were not uncommon on peakism's online forums, blogs, and comments sections. Almost paraphrasing Earle, "Cashmere" wrote that "post peak will be a reversion to conventional norms, like 'em or not."[10] Others were far less reserved about their views on shifting gender roles behind the cloak of anonymity. "Bshirt," for example, predicted: "All those braindead female public school teachers making their 50–70 thousand dollars a year teaching seven year old kids to count to ten will be . . . damn glad to be in a kitchen with a roof and food in it. But with their attitude . . . it's highly doubtful the majority will be wanted by anybody." "Mos6507" alleged that post-apocalyptic survival won't be "like sharing chores": "It's a matter of life and death. Political correctness goes out the window and he (or she) who demonstrates better competence as a warrior gets dibs. What that means is Ripley [from the film *Alien*] comes before [Steve] [U]rkel [the iconic nerd from the television show *Family Matters*] and Rambo [from the *Rambo* films] comes before Paris Hilton." "Novus" attempted to add some historical context to these claims: "Feminism was a failed experiment partly enabled by cheap energy and a society with excess wealth. The reality we inherit contains neither cheap energy or excess wealth among other factors which will lead to a return to patriarchy."[11]

Some female peakists agreed that gender roles might change. Gail Tverberg, who authored a column as "Gail the Actuary" on The Oil Drum, granted that, "to the extent that physical labor becomes more highly valued" in a somatic energy regime, "women may see their status go down."[12] "Ayame" predicted that "there will be a return to moderate patriarchy out of necessity to protect the family and clan group." She continued: "As a women [*sic*] I would actually welcome the protection of a male clique in return for doing monotonous tasks. Generally men are better genetically adapted to violence both physically and psychologically. Sure women might not have all the opportunities they used to but when the group is threatened the men will [be] the ones going out and sacrificing their lives for their families."[13]

Race and Peakism

While the lack of women in the peak oil movement was a regular topic of conversation, the absence of people of color was rarely debated. Fully 89–91 percent of my respondents self-identified as "white" or "Caucasian," while only 72.4 percent of Americans identified themselves as such at the time of the surveys.[14] If anything, my informal canvasses of audiences at peak oil conferences and lectures showed fewer, not more, people of color. Potential explanations for this absence might begin not just with the open historical exclusion of people of color from participating in public activism before the civil rights movement but with the history of the exclusion of people of color from environmental movements specifically. Communities of color in the United States are and have always been the primary victims of industrial pollution, chemical toxins, and the general distribution of environmental risk (such as lead poisoning, waste treatment siting, air pollution, and pesticides),[15] but until recently these issues were considered *urban* and not *environmental*. Many environmental issues are still framed as universal problems that affect all Americans (or humans) equally, without regard to the dimensions of race, class, and gender that often leave disadvantaged groups to bear the worst burdens. While the stereotypic profile of environmentalists (white and well-off) has certainly changed in the last two decades and environmentalists and scholars now recognize the concerns and campaigns of people of color as environmental even if they are not publicly framed as such, this history of exclusion undoubtedly leads some people of color to avoid explicitly environmental concerns and actions.

That said, the period 1990–2010 saw an extraordinary growth in the number of groups organizing around "environmental justice" in the United States, many of them founded and led by people of color.[16] These groups actively opposed and protested environmental discrimination throughout the country by utilizing all the forms of politics available to them, including lobbying, holding public meetings, providing "toxic tours" of affected neighborhoods, organizing public protests, and engaging in direct action. Given this history of exclusion and vulnerability to and awareness of environmental discrimination, it is perhaps not surprising that the environmental efforts of many people of color have focused on more immediate issues instead of less tangible concerns about the future, such as energy depletion.[17]

But, instead of wondering why people of color did not subscribe to peakism, we might ask why so many middle-aged white Americans did. Although race was rarely even mentioned on peakist forums (owing, perhaps, to the possible presence of peakists of color and a liberal disapproval of open racism), some visions of the post-peak world clearly tied the revitalization of masculinity to whiteness. In *World Made by Hand*, for example, characters of color are invisible, an absent presence involved in spatiotemporally distant race wars. Since there are no such conflicts in Union Grove, we are left to conclude that they must be at least partially responsible for them.

In other peakist speculations, the connection between whiteness and masculinity is far more explicit. In W. R. Flynn's *Shut Down* (2011), the peaceful white community that assembles in Corbett, Oregon, after the peak oil collapse survives only by defending itself in a race war against marauding gangs of African Americans and Latinos from urban Portland. These "armed, drunk, stoned" gangs, "dressed similarly in oversized t-shirts and low-hung baggy pants," commit acts of theft, sexual violence, and torture throughout the city. The novel's protagonists join a group of peaceable whites, such as the blue-collar "Big Don, an energetic, charismatic and strapping 6' 1", 190 lb, 72-year old, and tough as nails retired railroad worker" who wears "faded jeans and a long-sleeved flannel shirt." As noted in the last chapter, mainstream post-apocalyptic fantasies generally present a vision of multicultural harmony, but Flynn's does not. Even "migrant" Latinos, despite having "worked [in Corbett] on the same farms for years" and thus potentially possessing useful knowledge about local agriculture, are said to "pose an unacceptable level of resource burden and were being told to leave or turned away if they were encountered by a patrol."[18]

The regressive gender and racial formations of some peakist visions and their articulation to a socially conservative message will be familiar to devoted observers and consumers of post-apocalyptic popular culture. To recall the political connotations of the pop cultural postoil dystopia, we might look back at two nondisaster films that had a clear impact on peakists and peak oil authors, *Mad Max* (George Miller, 1979) and *Mad Max 2: Road Warrior* (George Miller, 1981). Over three-quarters of all respondents had seen each film, and peakists referred to both regularly in their responses—a female adherent, for example, said that "the *Mad Max* scenario always springs to mind" when she thinks of the future, while a Connecticut man reported that "some people wish they lived in a mad max world."[19]

Released in 1979 and 1981 in Australia and soon thereafter in the

United States, the *Mad Max* films caused a sensation stateside. Although their landscape and playful absurdity were quintessentially Australian, their director, George Miller, seamlessly adopted elements of the Hollywood western genre. Set in the near future, *Mad Max* presents a postnuclear, deserted, Hobbesian landscape where social and political institutions are rapidly disappearing. Only the heroic actions of "the Bronze," leather-clad policemen such as protagonist Max (Mel Gibson), prevent gangs of drug-addled motorcycle outlaws from committing even more thefts, rapes, and murders. In the second film, set some years in the distant future, a now solitary Max—his wife and child murdered in the first film—aids a small community (a "civilized society") struggling to guard their oil rig compound and treasure trove of petroleum against a depraved band of desperate outsiders. Here, Max is the stranger come to town, described as the "Man with No Name" in an explicit reference to spaghetti westerns.

The *Mad Max* films struck a chord with American viewers for a number of reasons. The films were topical as oil scarcity and energy independence were regular concerns for Americans in the 1970s. Remarking on the connection between *Mad Max* and petroleum dependency, the screenplay coauthor, James McCausland, said in 2006: "George [Miller] and I wrote the script based on the [seemingly accurate] thesis that people would do almost anything to keep vehicles moving and the assumption that nations would not consider the huge costs of providing infrastructure for alternative energy until it was too late."[20] But, if the series was intended as a liberal cautionary tale about fossil fuel dependency, in the early 1980s it also mirrored the law-and-order politics of conservatives such as Richard Nixon and Ronald Reagan.[21] In the first film, Max arrests a marauder and brings him to court, only to find that the witnesses are too intimidated to testify. Liberal law enforcement solutions fail to bring the killer to justice, and he walks free. Like the heroes of similar action films of the era, such as the *Dirty Harry* (1971–88) and *Death Wish* (1974–94) series, the white male hero here can enact justice only by stepping outside the law.[22] As in *World Made by Hand*, the post-apocalyptic crisis brings about a return to "traditional" (patriarchal) gender roles, consistent with Reagan's "family values" rhetoric and evocation of the citizen as a solitary, disembodied actor.[23]

These conservative tropes were emphasized by the threat of outsiders in both films. In the 1980s, Reagan and other conservatives regularly asserted that various minority groups (such as African Americans, leftists, and homosexuals) posed a threat to the "traditional" American social order.[24] In *Mad Max* (and some peakist fantasies), villains are coded to embody the threats posed by each of these groups, post-apocalyptic punks adorned with leather clothes, tattoos, earrings, and mohawks. Although punks are not necessarily leftists, they represent the threat of youthful rebellion that has been associated with the Left since the 1960s. Especially in *The Road Warrior* (1981), the villains also symbolize the threat of deviant sexuality. Their leader, Humungus, is nearly naked, with a body-builder physique, crisscrossing leather straps, and a hockey mask, and two male followers form a gay couple: one sports a pink Mohawk and wears feathers around his neck, and the other has long blond hair and a black leather top with nipple cutouts. In both films, the threat of nonnormative sexuality is not merely hypothetical since the outsiders are sadists who gang-rape their victims.

At the same time, Humungus and his followers signify the racial other via the film's connection to the "savages" of bygone Hollywood westerns, Native Americans. They are adorned with feathers, face paint, and mohawks, and they circle the "fort" of the community's oil rig in souped-up motorcycles. Similarly, in *World Made by Hand*, Wayne Karp and his gang sport a *Mad Max* punk aesthetic, are racialized outsiders who are described as "less like their own parents and forbears and more like the Iroquois who had inhabited the same area four hundred years earlier," and are sexual deviants who brutally sodomize the town preacher when they take him captive.[25] In *Shut Down* (2011), the black and Latino gangs of "sadists" pose an equally phantasmagoric sexual threat to heterosexual whites, declaring that their march to Corbett will provide "cracka' hoe's fo all a bro's" as they dance "in the middle of the street inches away" from one another, "eagerly bumping and grinding away on each other."[26] In these ways, some post-peak visions walked the socially conservative and often racist path of yesterday's post-apocalyptic fantasies and nightmares.

White American Masculinity in the Twenty-First Century

The similar articulations between whiteness, masculinity, and heteronormativity in *Mad Max*, which found a cult audience in the United States in the 1980s, and peakists' predictions in the first decade of the twenty-first century might suggest that little had changed in the intervening three decades. But while some peak oil visions of a reinvigorated white masculinity clearly drew on *Mad Max* and other sources, we should see them as a response to the particular concerns and anxieties around white masculinity in the twenty-first century. Of particular importance are changes in American political economy, the anti-immigrant "Latino threat" narrative, and fears of reverse gender discrimination in the new global economy.[27]

While there is good reason to be skeptical of the latest declaration of a crisis in masculinity, there have been undeniable shifts in labor and political economy that transformed life for many American men over the last half century. Deindustrialization has been under way since the early 1970s, of course, but the decline in manufacturing employment as a percentage of total work has been particularly sharp over the last decade. While the total number of manufacturing jobs in the United States has remained relatively stable for the last forty years, the nation's population has increased, with the result that manufacturing workers now make up less than 10 percent of all workers, down from 20 percent in 1980 and nearly 30 percent in 1960.[28] Whereas these blue-collar jobs once promised a middle-class salary and benefits—as a result of hard-fought early to mid-twentieth-century union victories—that is less and less the case. Shifting patterns in employment and politics led to the transfer of large numbers of jobs from union-heavy Northeast and Mideast states to right-to-work states in the South, while changes in labor law made unionization more difficult.

Furthermore, the actual work of blue-collar employment is not what it once was. Jobs that are considered manufacturing are now often quite similar to what we usually think of as white-collar enterprises. Owing to new technologies and automation that increased productivity (and rendered many workers unnecessary), factories are now as likely to be called *technology distribution centers* and require workers that are well educated and computer literate.[29] According to one analysis, three of ten American jobs counted as manufacturing are "things that would look to most people like white-collar service jobs: Sales, engineering, design, that sort of thing."[30]

The psychological influence of the shift from blue-collar to white-collar work was a frequent topic of conversation in the late 1990s, when groups like the Promise Keepers, a Christian men's organization, and events like the Million Man March sought to redefine masculinity. In popular culture, where dominant gender roles and expectations are often forged or expressed, a slate of films, such as *In the Company of Men* (Neil LaBute 1997), *American Beauty* (Sam Mendes, 1999), *Office Space* (Mike Judge, 1999), and especially *Fight Club* (David Fincher, 1999) dramatized the alienation and repressed anger that heterosexual white men were said to be suffering as a result of the cubicles and corporate culture of white-collar work.[31] In her influential *Stiffed* (1999), Susan

Faludi observed that the nation's "pulse takers," including "newspaper editors, TV pundits, fundamentalist preachers, marketeers, [and] legislators," agreed that "American manhood was under siege" in the 1990s.[32] As one angry white man, *Fight Club*'s Tyler Durden (Brad Pitt), put it, some (white) American men saw themselves as "an entire generation pumping gas, waiting tables," "slaves with white collars" without "purpose or place": "We have no Great War, no Great Depression. Our great war is a spiritual war. Our great depression is our lives."

Such claims should be taken with a grain of salt since there was hardly a decade in the twentieth century that masculinity (meaning white masculinity) was not declared to be in a state of crisis by one constituency or another. In the Progressive Era, as Kevin P. Murphy has shown, Theodore Roosevelt and others promoted "strenuous manhood" by contrasting the "redblood" American with the "mollycoddle," the all-too-common "weakling and the coward" who were "out of place in a strong and free community."[33] In the 1950s, as K. A. Cuordileone has documented, Cold War masculinity was threatened by "corporatism and the decline of the self-made man; the affects of affluence and comfort; 'civilizing,' emasculating women; the power of a sentimental, feminine mass culture; and the excessive influence of women on boys and men."[34] Responses to second-wave feminism in the 1970s and 1980s, as Michal A. Messner has observed, included openly antifeminist "men's rights" organizations such as the Coalition of Free Men, which claimed that "it is actually women who have the power and men who are most oppressed."[35]

However, as Faludi and others have noted, men's roles in the twentieth-century United States became closely tied to their labor, and there were major political economic developments since the 1970s that changed men's work roles, owing primarily to the decline of unions and neoliberal globalization. By 2010, when media commentators returned to the subject of masculinity in crisis, these trends had only accelerated. Between 2000 and 2010, the United States lost nearly one-third of its manufacturing jobs as over forty thousand factories closed. The economic recession of 2008 had a disproportionate impact on men—according to one estimate, they held nearly three of every four jobs that disappeared between 2007 and 2011. Indeed, men's unemployment rate during this period exceeded women's by just 25 percent only because so many men stopped actively searching for employment and, thus, were not counted by the Department of Labor's official statistics. According to some journalists, the future looked even more dire. Some surmised that changes in the global economy, which had supplanted American manufacturing with automation or low-wage work in the Global South, might very well signify "the end of [American] men" since "the vast majority" of "the 15.3 million new jobs projected to sprout up over the next decade will come in fields that currently attract far more women than men."[36] Others noted that women now earned 60 percent of bachelor's and master's degrees, increasingly a prerequisite for steady employment beyond low-wage, low- (or no-) benefit service jobs.

Even so, these declarations were premature, a fact that many journalists mentioned below the fold, if at all. In 2011 only 4 percent of Fortune 500 CEOs and 17 percent of US senators were female. Women earned, on average, 17.8 percent less for performing the same work as men and received unequal pay in nearly every field—women with the same education and experience earned less than men in 527 of the 534 occupations listed by the Bureau of Labor Statistics.[37] While baby steps were certainly being taken toward gender equity, talk of true equality, let alone the "end of men," was and is unwarranted. Discourse about a crisis of masculinity, then, was as much about perception as it was about reality, as it has always been. But globalization and deindustrialization had real psychological consequences—one peakist surmised that, in losing his job during the recession, his father had "lost the former relevancy that gave him the opportunity to make a living": "He is suffering."[38] In a postindustrial society, as Faludi observed, many men not only lost the "utilitarian world" of blue-collar labor that once provided steady employment but were also "thrust into an ornamental realm" where manhood was now defined by "appearance, by youth and attractiveness, by money and aggression, by posture and swagger and 'props,' by the curled lip and petulant sulk and flexed biceps."[39] Predicting the reversal of this ornamentalization, one peakist, "Bigdoug2053," claimed that "things will be better for succeeding generations" in the post-peak world "as localized economies eventually give men meaningful, though likely physically harder, work that will help support families and communities."[40]

In the first decade of the twenty-first century, this ornamentalization resulted in the declaration of a new masculine identity: the metrosexual. As Toby Miller put it, metrosexual men were "feminized males who blur the visual style of straight and gay in a restless search 'to spend, shop and deep-condition.'"[41] While most men have always been implicitly aware of the dominant conceptions of the ideal male body, fashion, and certain aspects of home decor, metrosexuality required a new level of knowledge and curation of the self, primarily through mass consumption. The growing acceptance of homosexuality, at least in blue states and metropolitan centers, led to blurred lines between the public and the private standards of gay subcultures and straight men, with the television show *Queer Eye for the Straight Guy* (2003–7) as a prominent example.[42] By the time of the economic recession, the expectations for even heterosexual, nonmetrosexual American men seemed to be changing, with the explosion of new skin creams, false eyelashes, nail polish, grooming tools (for "manscaping"), and hair care products designed specifically for men.[43]

Retrosexuality: Back to the Future

Even as the aforementioned shifts in political economy, culture, and gender roles were creating the need for a new conception of white male heterosexuality—or, from a more utopian standpoint, the elimination altogether of categories of identity based on essentialist notions such as whiteness and masculinity—popular conceptions of masculinity were looking backward. This *retrosexuality*, as some observers called it, began in the wake of the 9/11 attacks.[44] President George W. Bush's "dead or alive" cowboy machismo, though mocked by liberals, provided a familiar model of masculinity that sought security not in an acknowledgment of a changing twenty-first century but in a mythologized American history of western individualism and mastery over nature.[45] In the same vein, the counterterrorism expert, epitomized by Jack Bauer in the television show *24* (2001–10), harked back to the cowboy as a model of heroic masculinity. Bauer, a patriot and a man's man, is willing to use violence and step outside the laws of society in order to protect it. No less an authority on sexuality than former president Bill Clinton praised Bauer, once telling NBC's *Meet the Press* that, "when Bauer goes out there on his own and is prepared to live with the consequences, it always seems to work better."[46]

In a cultural moment that seemed to celebrate obstinacy and rigidity, other noteworthy models of masculinity were similarly backward looking. Firemen in particular, lionized after the Fire Department of the City of New York's fatal rush into the World Trade Center towers, "came to stand as exemplars of a way of life that was encoded as particularly and typically American: selfless, heroic, individually brave citizens rising in a moment of need to serve the common good," as Hamilton Carroll observed.[47] As a bastion of blue-collar white ethnic masculinity (and even segregation) during a moment rife with anti-immigration sentiment and a unionized public worker during a neo-liberal era, the fireman was a particularly anachronistic model of hero-ism in the early twenty-first century.[48] The glorification of disappearing blue-collar work in popular culture, on television shows such as *American Chopper* (2003–10), *Dirty Jobs* (2005–12), *Deadliest Catch* (2005–), and *Ice Road Truckers* (2007–), followed the heroic mold of the 9/11 fire-fighters: backward, not forward.

Of course, any analysis of early twenty-first-century heterosexual masculinity must acknowledge that many of the dominant models of masculinity in popular culture of the period did not fit into a retrosex-ual model and often featured people of color. Some of the most visible and admired male trendsetters during this period were black hip-hop artists such as Jay-Z and Dr. Dre, movie stars such as Will Smith and Samuel L. Jackson, and athletes such as Dwayne Wade, Kobe Bryant, and LeBron James. At the same time, the lifestyles of these celebrities can be viewed as beyond the reach of the average white male because their success can be (mis)interpreted as tied to their race—in the case of athletes, their "natural" athletic ability; in the case of musicians and movie stars, their "natural" cool.[49] In contrast, the kinds of back-to-the-future masculinities I highlight in this chapter are freely available to all. To some extent, we might see the focus on regular guys such as the fireman and the blue-collar worker as a response to the media's lioniza-tion of wealthy athletes, musicians, and movie stars.

When the term *retrosexual* was coined, it referred to the strong, si-lent, charismatic Don Draper of the television show *Mad Men* (Mat-thew Weiner, AMC, 2007–15), but it quickly morphed into something broader: the "real men" of the Greatest Generation. As one cultural commentator put it: "[The retrosexual is the] anti-metrosexual, the opposite of that guy who emerged in the 1990s in all his pedicured, moussed-up, skinny-jeans glory. That man-boy was searching for his inner girl. . . . The retrosexual, however, wants to put the man back into manhood."[50] Blogs and Facebook pages devoted to the trend popped up, along with *The Retrosexual Manual: How to Be a Real Man* (2008), which advertised that "it's time to go back to basics—back to when men were men and women made breakfast the morning after."[51] One critical review of the term argued that it was "about returning to a time when men were men . . . [a] simpler time when it was clear what it meant to be a man and what his responsibilities were."[52] Proponents of retrosexuality, like the authors of The Art of Manliness blog (and book), claimed that relearning "Manly Skills," such as "How to Wire an Outlet," "How to Raise Backyard Chickens," and "How to Bug-In: What

You Need to Know to Survive a Grid-Down Disaster," were a necessary part of regaining one's manhood.[53]

If we recall the kinds of activities described in chapter 1, it should be clear that peakists, survivalists, and so-called retrosexuals were engaged in overlapping activities. While each group might be engaged for seemingly different reasons, I want to suggest that retrosexuals and some peakists were responding to the same impulse: a resistance to the perceived feminization of labor and the specialization of knowledge that, without alternate models of masculinity, looked backward instead of forward. As the white male peakist whose words are cited as one of this chapter's epigraphs put it: "I have roamed North America at will from Ixtapa to Halifax. I eat like a king. I want for nothing. I know that my ancestors were more connected to nature, that they were survivors, but I am wealthy and soft."[54] Retrosexuality, in all its forms, held the promise of making these men hard again.

Antifeminism and the Latino Threat

That the bygone era when men were men happened to coincide with a period of enormous white male privilege, before feminism and the civil rights movement, is no coincidence. However unpopular antifeminist and anti-immigration political positions were among liberals and leftists (such as most peakists) throughout the period 2000–2010, they undergirded the popular models of white masculinity that emerged—the cowboy, the fireman, the counterterrorism agent, the Wild West lawman, and even the retrosexual prepper. Given the changing economic circumstances of the United States, what is required, as Andrew Romano and Tony Dokoupil argued, "is not a reconnection with the past but a liberation from it; not a revival of the old role but an expansion of it."[55] However, as Stephanie Coontz has pointed out: "One thing standing in the way of further progress for many men is the same obstacle that held women back for so long: overinvestment in their gender identity instead of their individual personhood." This overinvestment is not mere insecurity but is reified by cultural norms and economic incentives. As Coontz observed, statistics show that "men who take an active role in child care and housework at home are more likely than other men to be harassed at work" and that "men who request family leave are often viewed as weak or uncompetitive and face a greater risk of being demoted or downsized."[56] Similarly, the journalist Hanna Rosin noted that when she "asked several businesswomen in Alexander

City [Alabama] if they would hire a man to be a secretary or a receptionist or a nurse . . . many of them just laughed."[57] Since mainstream cultural models of alternate masculinity have not emerged, some white men—whether blue-collar workers who suddenly found themselves unemployed or white-collar workers whose masculinity seems merely ornamental—felt a sense of victimhood, caught between a changing economy and rigid gender roles.

This sense of male victimization was opportunistically yoked to xenophobia and sexism by conservative activists and demagogues. Leo R. Chavez has described the wave of xenophobic discourse and action in the period 1990–2010 as a "Latino threat narrative" that sees Latinos (particularly Mexicans) as "an invading source from south of the border" bent on "destroying the American way of life."[58] A 2008 American National Election Studies survey found that 44 percent of Americans found it "very" or "extremely" likely that "recent immigration levels [would] take jobs away from people already" in the United States, with another 41 percent considering it "somewhat likely."[59] In reality, there is little connection between job loss and increased immigration—most immigrants from Latin America, especially undocumented immigrants, are forced into working low-paying, no-benefit jobs that most whites have been unwilling to accept, even when the only alternative is unemployment.[60] Nonetheless, there are demographic and political changes on the horizon. Jonathan Chait noted that, as a result of immigration and, to a lesser extent, differentials in birth rates, "the nonwhite proportion of the electorate grows by about half a percentage point" ever year, with the result that, "in every presidential election, the minority share of the vote increases by 2 percent." Projecting these changes into the future, this trend suggests that, "in 30 years, nonwhites will outnumber whites" as voters.[61] Given that people of color, while hardly a monolithic voting bloc, have historically tended toward the Democratic Party—partially as a result of indifference and hostility from the Republican Party—the Latino threat narrative bears some relation to the future of electoral politics in the United States, if we squint hard enough.

We might not expect peakists and Tea Partiers to mix, and, indeed, their politics are, for the most part, diametrically opposed. But both groups are predominately white, male, educated, and middle-aged and predict an impending decline on the horizon.[62] While many male peakists did not fit the model of the retrosexual, the comments of some male adherents on Internet forums make it clear that conservative articulations of male victimhood, tied to antifeminism and the Latino threat narrative, have crossed some otherwise impermeable political boundaries.

Retrosexuality on the Small Screen: *Revolution*

The surprising political overtones of some post-peak futures are also implicit in many of the post-apocalyptic visions now prominent in American popular culture. A post-apocalyptic reimagination of the crisis of masculinity caused by the feminization of labor and the loss of blue-collar jobs during the recession is a central theme of the first season of the television show *Revolution* (Eric Kripke, NBC, 2012–14), which received a great deal of hype in 2012, in part it because it was produced by J. J. Abrams (of the television show *Lost*) and directed by John Favreau (of the *Iron Man* film franchise).[63] Although hardly exceptional in its premise or execution, the show is relevant here because of the way that it reflects common themes in current and recent American popular culture.

Set fifteen years after "the blackout," when all electronics and engines suddenly ceased to function, *Revolution* presents a familiar Hobbesian post-apocalyptic terrain. The former United States is controlled by a patchwork of regional militias, while the rest of the survivors eke out a living through subsistence farming. Although the show's nominal star is the teenage daughter, Charlie (Tracy Spiridakos), her character is an anodyne blend of goodwilled naïveté and kick-ass girl power. As is often the case, Charlie is simultaneously a strong woman who shows that women are just as capable of violence as men (the bow being the current weapon of choice) and heavily sexualized—a number of television critics objected to her skimpy outfits, plentiful makeup, and perfectly coiffed hair (see fig. 11).[64] The show's real drama revolves around the battle of a group of men for power in the post-apocalyptic world, with regular flashbacks to the contemporary United States that serve to naturalize existing formations of masculinity.

The shortest-lived of these male characters is Ben Matheson, Charlie's father, who is killed by the Monroe Militia in the first episode. A caring father who worked at the University of Chicago as a scientist developing renewable energy sources, Ben holds values that are out of step with the world that the blackout creates—or returns humankind to. In a flashback, he is shown leading his family out of Chicago until a scavenger steals their wagon of food. Holding a shotgun to the man as he absconds with their provisions, Ben finds himself unable to fire,

FIGURE 11 NBC's *Revolution* presents a post-apocalyptic world in which masculine individualism and the capacity for violence and physical labor are rewarded. *Far left*: Former Google millionaire Aaron Pittman. *Center left*: Teenager Charlie Matheson. *Center right*: Former soldier Miles Matheson. *Far right*: Former white-collar worker Tom Neville. *Source*: http://www.nbc.com.

and his wife must step forward to accomplish the task. In the show's present, the Monroe Militia arrives at the family's commune to capture Ben. Seeking to negotiate instead of defend himself—as, it is suggested, he had done many times before, when he surrendered weapons and food stores—he is ultimately shot to death and, thus, unable to prevent his son being taken prisoner.

With his last words Ben sends his compatriots to seek out his brother Miles in Chicago. They find a man tough and hardened, drinking a single-malt scotch while awaiting what appears to be certain death at the hands of the approaching militia. But Miles turns out to be the former commander of the militia and a military-trained assassin. With an eternal five-o'clock shadow and a flip remark for every situation, he is a classic rogue, but he is also able to easily dispatch a dozen men with a sword or his bare hands, and his training in the preblackout US military has equipped him for success in any situation. In the post-apocalypse, as in our current world, the show suggests, military might and the ability to inflict violence on others is the most important survival skill of all.

Revolution's perspective on the feminization of labor is highlighted by the contrast between Aaron Pittman, a friend of the late Ben Matheson, and Tom Neville, a major in the Monroe Militia. Before the blackout Aaron was the "wizard of Google," with a beautiful wife, four homes, a limousine, a plane, and three hundred employees working under him. When we first meet him after the blackout, however, he is clearly a burden on his community. His weight and physical incompetence—he is chubby and wears thick-rimmed glasses—are regular sources of humor. He insists on joining the mission to recover the captured Danny but admits that he is "afraid of bees," complains about "chafing" issues, and acknowledges that he is often "weak and afraid." Considered a genius before the blackout, he is revealed to be less than a man in the post-apocalyptic world. Months after the blackout, he and his wife deserted the city, but she contracted dysentery after he allowed her to drink from a sewage-filled lake. Soon thereafter, he failed to protect her from two looters, and they were saved only by the aid of a stronger, leaner man who excels in physical combat. Aaron abandoned his wife the next day, believing that she would fare better under the protection of a group of strangers.

Tom Neville, once a white-collar worker, undergoes the opposite transformation. On the day of the blackout he was berated and then fired by a younger colleague for granting an insurance claim that he could have denied, his hands fidgeting as he sat in an office chair in a button-down-collar shirt and a tie. When he returned home, he politely asked his neighbors to turn their music down but was ignored. He began his transformation when the same neighbor broke into his home six weeks after the blackout: Neville pummeled him to death and rose a new man. When we see him fifteen years later, he is a leader of the militia: competent, strong, and ruthless.

When Neville holds Aaron and Charlie hostage, he delivers a monologue that encapsulates the show's view of post-apocalyptic masculinity: "I know you. I recognized you the minute I laid eyes. That's Aaron Pittman, the Wizard of Google. You been on the cover of *Wired* magazine more times than I can count. I bet you were high and mighty when the lights were on! I bet you'd boss around those poor bastards in their tiny cubicles! I was one of those poor bastards myself. But now look at you, and look at me. Now you need Miles saving your fat, pockmarked ass!"[65] This diatribe combines a number of ideas that were common in American politics and popular culture at the time. Neville resents the power that elites like Pittman held over middle-level workers like Neville himself and learned from the blackout that stereotypically liberal qualities, such as empathy, compassion, and rational discourse, are no longer rewarded—if they ever were. Concern for others, it is emphasized in almost every episode, is a potential weakness. Most importantly, in positing a Hobbesian scenario of all against all in a post-apocalyptic world that is often confused with a state of nature, *Revolution* naturalizes violent, patriarchal gender constructs.

Race, Gender, and Individualism

If, as discussed in chapter 3, the Internet of the 1990s and the early years of the twenty-first century is one ideal version of libertarian individualism, the post-apocalyptic scenario is another. Claire P. Curtis has written that, since post-apocalyptic narratives speak to "our desire to start over again," they dramatize the "state of nature" that political scientists have theorized about for centuries.[66] While Curtis's analysis gives equal attention to Locke, Rousseau, Rawls, and Hobbes, it is really the latter's violent vision of total individualism and lawlessness that dominates the genre in most of its recent incarnations. This holds true whether we are focusing on peak oil fiction, disaster films, or the remarkable renaissance of novels, films, and games about zombie-attack scenarios, which usually constitute "the apocalyptic invasion of our world by hordes of cannibalistic, contagious, and animated corpses."[67]

Theoretically, a certain line of thinking goes, a Hobbesian world in which government and social structures such as race, class, and gender break down would provide complete equality of opportunity, which would allow the strong, meritorious individual to persevere and survive. The individual could finally become what he should have been already had he not been held back by the strictures of modern civilization. Elizabeth Fox-Genovese noted that individualism (as a political philosophy) has "tended to depersonalize the individual, which it represented as simply a unit of society and the polity." Indeed, to many, individualism's appeal is its apparent gender- and color-blindness: the individual "might be male or female, tall or short, rich or poor, nurturing or aggressive, for such attributes did not affect the individual's status as an individual."[68]

Although it flashes the promise of egalitarianism, individualism is, we should recall, a distinctly raced and gendered philosophy. First, history has a long shadow. Until relatively recently, women, people of color, gays and lesbians, the poor or unpropertied, and other groups were systematically oppressed, disenfranchised, and/or enslaved, and this treatment has consequences that reverberate for decades, if not centuries.[69] In 2015, these groups are still routinely denied opportunities and subject to institutional prejudice and individual acts of discrimination. Second, individualism relies on a theoretical equality of opportunity that has never existed in fact. To the extent that it ever exists, true individual freedom has been enjoyed in the United States only by white, Christian, wealthy and healthy heterosexual men. It is for this reason that scholars such as Carol Pateman and Charles W. Mills have critiqued the idea of a universal social contract as being a distinctly gendered and raced concept.[70]

For those who refuse to recognize or acknowledge this fact, individualism can easily lead to prejudice and resentment. As Jack Turner observed: "At individualism's heart is a will to see the world in a self-congratulatory way, to construe one's achievements as entirely self-authored, to interpret accidental privileges as just deserts." In this way, "the individualist's vainglory gives rise to distorted social perception" about the apparent inferiority of other groups.[71] Absent an acknowledgment of historically persistent structural racism, sexism, and classism and continuing discrimination, individualists find themselves justifying a given group's poverty, powerlessness, low wages, or rate of incarceration by referring to the individual failures of its members. When this "distorted social perception" is applied to topics such as federal attempts to reverse centuries of discriminatory practices, it can breed racial resentment and claims of reverse racism. As Sally Robinson put it, "those who defend the existence of an unmarked, universally available individualism," whether in the present or the post-apocalyptic future, are themselves practicing "identity politics in both subtle and overt ways."[72]

The link that I have drawn between peakism and socially conservative individualism does not hold true for all peak oil believers, of course—some male commenters objected to the sexist and xenophobic connotations of *World Made by Hand*, for example. Nor, as I will show in the conclusion, does it render their sense of an impending environmental crisis illegitimate or invalid. But the connections between post-apocalyptic imaginaries and the shifts in political economy, gender roles, and racial scapegoating in the United States at the turn of the millennium provide a compelling explanation for the whiteness and maleness of this population. We might note that the kinds of apocalyptic scenarios that peak oil authors and believers often focused on—driven by scarcity and peopled by individualists—can lead them to adopt positions that seem counter to their stated beliefs. While the retrosexual activities that peakism and related ideologies encourage are not in themselves problematic—raising chickens and learning to farm are harmless, if not praiseworthy, endeavors—we should always remain mindful of the political, racial, and sexual connotations that they carry in their wake.

That said, we might also recognize the difficult project that peak oil believers were engaged in: imagining life after oil. If we recall the history related in chapter 2, we might note that the social roles that Americans grew accustomed to inhabiting over the last century were (and are) dependent, directly or indirectly, on the incredible energy subsidy of fossil fuels. Life after oil—whether the transition is pushed by climate change, energy depletion, or (likely) both—will, thus, require not only an economic, technological, and infrastructural but also a social and cultural transformation that is beyond most of our imaginations. If some peakists' visions were biased by anxieties, fears, and hopes that were surprisingly common in the early twenty-first-century United States, they might be pardoned for the intimidating scale of their endeavor: imagining and cultivating a post-carbon world that does not merely replicate the injustices of the past.[73]

As I examine in the conclusion, this project might have seemed futuristic and fanciful a decade ago, but by the time of this writing it was increasingly imperative.

Climate Change
and the Big Picture

The surveys and ethnographic research for this book were conducted from 2007 to 2011, but by the time it went to press Americans seemed to be facing a very different energy landscape. After rising by over 300 percent between 2000 and 2009, the average price of a gallon of gas in the United States had leveled off.[1] While global energy consumption had grown by 15 percent between 2005 and 2011, global conventional oil production had increased by only 4 percent during this time, unconventional petroleum sources having made up the difference.[2] The whispered fears of depletion that gave rise to peakism had been replaced by a hearty cornucopianism fueled by the boom in American natural gas (via hydrofracking) and Canadian tar sands. Although both these sources had a much lower energy returned on energy invested (EROI) and are (in different ways) potentially more damaging to the environment, the rising price of conventional oil production made them quite profitable.[3] Once again, the petroapocalypse was postponed.

As a result of these developments, general interest in the topic of energy depletion waned. There were fewer new converts to be found, once-vibrant forums withered, and the central Web site The Oil Drum shut its doors in 2013.[4] By 2012, pundits and bloggers triumphantly announced that the peak oil movement had itself "peaked."[5] CNN's David Frum reported that, instead of a "long steady decline" of oil production, the United States would achieve energy self-sufficiency by 2030, and *Forbes*'s David Blackmon wrote that the "'Peak Oil' theory has basically gone the way of the California Condor, from widespread existence and acceptance in the oil and gas environment to near extinction . . . thanks to the discovery of and ability to access massive oil shale reservoirs not just in the United States, but all over the world."[6] These new resources are themselves finite and limited, of course, but energy insiders were already looking toward new frontiers, such as methane hydrates (deposits of frozen methane deep beneath the sea floor).[7] By 2014, most interested observers no longer believed that we would soon run out of hydrocarbons but were increasingly concerned that we might ultimately be drowned as a result of our consumption of them.

Where Are They Now?

In July 2013, I invited respondents to my first survey to answer a new set of questions about how their interest in resource depletion, climate change, and other issues had developed in the intervening two years. How did they think the energy and environmental landscape had changed? What were they thinking of doing differently, and what was the same? How had their dedication to peakism affected their lives? (Interested readers can see a list of questions in appendix 2.) In some ways their answers surprised me. Less than 10 percent of those who responded had significantly questioned their dedication to peakism, and the vast majority stood firm in their convictions and life course.

Although one believer admitted that he "thought things would be worse by now," nine of ten respondents said that their views had not changed. They were unimpressed with the potential of tar sands and natural gas, which they expected to merely postpone the expected collapse a decade or so. Conventional petroleum, they said, "was the cheap stuff, and it's in decline." As a female attorney from California put it: "The data behind peak oil and fossil fuel depletion has not changed. The trajectory is as bleak as ever. What has changed is the rhetoric surrounding the topic." That rhetoric now highlights unlimited futures instead of impending scarcity. Or, as a white male in his midseventies put it: "Nothing has really changed except the price of oil has enabled more expensive oil to be produced." A physician in her fifties likened this delay to "clacking along at the top of the roller coaster, just waiting for the first drop": "The fracking boom is prolonging the bumpy plateau at the peak."[8]

According to many, the delay was particularly problematic because fears of depletion had been a powerful impetus for a transition away from our fossil fuel–dominated energy regime sooner rather than later. A white male in his late forties believed: "The effect [of the new sources and methods] will be only to delay the inevitable and desensitize most people to the reality of the situation. In other words, to make it harder to adjust and cope when the crunch finally comes." A transcriptionist said: "If anything it will only create a sense of false security at a time when action is needed both to develop renewables and address global warming." Indeed, many had shifted their focus from oil depletion to climate change. A New York man reported: "Although I still follow energy issues very closely I have diversified my focus and am much more concerned about climate change and . . . the now increasingly apparent effects of changes that are occurring in our environment." A Canadian in his sixties said: "The only real change is the realization that the efforts to squeeze out the last fossil fuels from tarsands [sic] and shale will cause massive environmental and social destruction on the way to peak fossil fuels, while also worsening climate change."[9]

The vast majority had persisted in the course of action that they had embarked on after "converting" to peakism. An engineer reported: that "At the time [of the last survey] . . . all I could do was 'awaking' to the fact the future might not look like the present with a coat of green paint. Now I spend less time on the Internet consulting forums but more time doing: gardening or cooking." Similarly, an insurance agent from Ohio said that he was "pretty convinced on the issue so [he doesn't] really need reminding or reinforcing" and is now "much more interested in learning traditional skills of hunting, gardening and homesteading skills than learning about peak oil," while a white male in his sixties said: "Having arrived at the conclusions I had as per your first survey, I am less focused on learning more and more focused on responding to the inevitable. Since we've moved to homesteading most of our energies are going into that in a healthy and non-frenetic way, . . . [which has been] a great and pleasurable change."[10]

In this vein, a Maryland manufacturer said that he was doing "nothing different," with his home now "100% solar powered" and an electric car in his garage, and a city planner in North Carolina reported: "Now that I own a house and yard, I am focused more on on-site efforts to conserve energy and material resources: household energy efficiency, re-use of items . . . composting, maintaining the lawn and garden with low impact to environment and without chemicals, planting shade trees, etc." Some had made major changes to their lives and lifestyles since the previous survey. A former office manager, for example, was "overseas in South America exploring alternatives to the energy intensive lifestyles that North Americans tend to take for granted": "What I'm finding is that it's actually possible to be quite happy consuming less of everything without compromising quality of life." A young Floridian had moved to Seattle to "pursue [his] education in a more walkable environment": "I'm now living in one of the most walkable neighborhoods in the country where my grocery store, church, school, restaurants, bank, and pretty much all other necessities are within walking distance or a five-minute bus ride. For years I've been wanting to live this way, and I've never been happier. My carbon footprint is now incredibly lower." A female physician was "taking advantage of the respite provided by shale gas and tight oil to continue developing [her] properties to prepare for less energy and higher repair costs in the future" while teaching her children "that a simpler world can still be very fun":

We go camping and they have to learn how to do all the basics, fire starting, cooking over a fire, surviving in the wilderness. They are learning how to make all the foods they like from scratch, even time-consuming things like cheese, sausage, and breads. All of them can handle horses and ride. They are learning how to make clothing and build things without power. Each plays a different musical instrument and we gather to listen. Surprising how many of their friends want to listen, also. On Earth Day each year, we do without energy (other than the refrigerator). It's a fun day, a celebration of the future.[11]

Others failed to muster this kind of outlook. A female attorney in California opined: "No amount of 'education' is going to compete with the public relations machine that benefits from the false hopes of growth 'to infinity and beyond.' I have no hope that this will change until the wheels fall off the bus." A teacher from Maine reasoned:

Up until recently I was under the impression that it was possible to prepare. That ended a few months ago when I finally put most of the rest of the puzzle together and discovered that near term human extinction is likely [owing to climate change] and it is sure to be plainly evident within the remainder of my lifetime (three more decades at the very outside I imagine). So, the question no longer appears to be "how to prepare?" but rather "how do I choose to live with this knowledge and ultimately how do I choose to die?"[12]

Although this sentiment might appear overly grim to some read-ers, it might also be seen as the kind of philosophical reorientation that many scientists and scholars suggest the era of climate change demands.[13] Similarly, a retired Californian said: "I still wake up every morning wondering if this is going to be the day that reality intrudes on the American (and world) economy." A Washingtonian said: "Even though I understand that the life I enjoy as a first world resident is de-pendent upon actions and systems that are destroying the planet, I still love every day that the lights come on."[14]

Taking Crisis Seriously

Given the recent changes in the global energy landscape and the mo-tivations of some white, male peakists that we witnessed in the last chapter, it would be easy to write off the peak oil movement as merely another chapter in America's long history of misguided millennial-ism. This is not the approach I have taken. The subject of this book is a social movement a mere half decade in the rear-view mirror, but each chapter has argued that, to truly understand peak oil believers, we need to contextualize their concerns and actions. I have done this by presenting a number of interrelated perspectives on the twentieth- and twenty-first-century United States: a history of beliefs about abundance and scarcity, a history of oil production and prices, a history of the libertarian shift, a history of the development of Internet technology, a history of American apocalyptic thought, and a very recent history of white masculinity. In the first part of this conclusion, I ask the reader to consider one final perspective: our place in global environmental history.

Most readers will be familiar with many of the changes in the natural environment that have occurred over the last thirty years, so a lengthy recapitulation is not necessary. Instead, I want to simply remind the reader of our current position. Although our sense of an ongoing envi-ronmental crisis has become, as Frederick Buell put it, a "way of life," we are living in a moment of almost indescribable peril as a result of our reliance on fossil fuels.[15] The atmospheric carbon dioxide concen-tration recently passed four hundred parts per million, the highest concentration in at least 800,000 years.[16] Increases in average tempera-ture and sea rise are only the most well-known consequences of this dramatic change, however. Weather patterns are shifting around the world, leading to instability, violence, and mass migration, especially in the Global South. Scholars estimate that climate change adversely affects three hundred million people and is responsible for 300,000 deaths each year.[17] Under current trends one-third of the planet's land will become desert by 2100, and the proportion in "extreme drought" will jump from the current 3 to 30 percent.[18] Extreme weather events do not just seem more common; they are more common. There were three times as many hydrological weather disasters from 2000 to 2009 as there were from 1980 to 1989.[19] Ocean ecosystems stand at the brink of collapse, with 90 percent of the large fish having disappeared from the oceans, primarily the result of ocean acidification and wasteful industrial fishing.[20] An average of up to 150 different species go extinct every day, and, if current trends continue, up to 30 percent of all known species—2.6 million different species—will become extinct by 2050.[21]

Most presentations of this and similar information tend to conclude with a reminder that it is not too late. Indeed, it is not—what we do today and tomorrow will always matter, and a radical shift in course is always possible. However, recent rates of consumption, pollution, and despoliation are not holding steady or declining; they are accelerating, as they have been doing since approximately 1950 (see fig. 12).[22] For example, in 2010 humans emitted roughly thirty-seven billion tons of carbon, one-quarter from the United States alone; with the increases in consumption and energy use per capita expected to occur outside the West, that number may jump as high as sixty billion tons by 2030. But climate stabilization, a critical aspect of true environmental sustainability, would require the entire world to reduce its overall emissions to the current poorest level—in the United States, this would mean a 95 percent reduction.[23]

These basic facts are most often euphemized in the United States, by those who acknowledge them, as an imperative to move toward sustainability. The term *sustainability* (like *crisis*) has been stretched thin by overuse and deliberate misuse, but we should recall from whence it came. Although it means different things to different people, *sustainability* is often defined as it was by the World Commission on Environment and Development in 1987, as "a process of change in which the exploitation of resources, the direction of investments, the orientation of technological development, and institutional change are all in harmony and enhance both current and future potential to meet human needs and aspirations."[24] The phrase *sustainable development* emerged in the 1990s as a convenient environmental paradigm that incorporated environmental protection and a limits-to-growth attention to

The 'Great Acceleration'

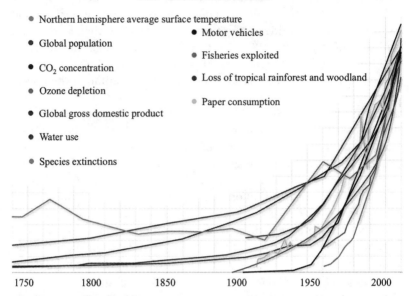

- Northern hemisphere average surface temperature
- Global population
- CO_2 concentration
- Ozone depletion
- Global gross domestic product
- Water use
- Species extinctions
- Motor vehicles
- Fisheries exploited
- Loss of tropical rainforest and woodland
- Paper consumption

1750   1800   1850   1900   1950   2000

FIGURE 12   While humans have always placed stress on the biosphere, these stresses have greatly accelerated since approximately 1950. Source: Garry Peterson, "Visualizing the Great Acceleration—Part II," Resilience Science, December 4, 2008, http://rs.resalliance .org/2008/12/04/visualizing-the-great-acceleration-part-ii. Data from William L. Steffen, Angelina Sanderson, Peter Tyson, Jill Jäger, Pamela A. Matson, Berrien Moore III, and Frank Oldfield, *Global Change and the Earth System: A Planet under Pressure* (Berlin: Springer, 2005).

resources and intergenerational equity with a corporate-friendly acceptance of prolonged economic growth.[25]

The very concept of sustainability is an implicit response to its opposite, unsustainability—a fair descriptor of contemporary human lifestyles, especially in the United States and throughout the Global North. One useful measure of sustainability is the ecological footprint, a conceptual tool that allows us to "estimate the resource consumption and waste assimilation requirements of a defined human population or economy in terms of a corresponding productive land area" on the basis of the biosphere's ability to regenerate.[26] As of 2007, human beings consumed the resources of one and a half Earths, but, if everyone on the planet lived as the average American does—as many people seek to do—we would exhaust the resources of five Earths each year.[27] The amount of resources and waste sinks our current way of life demands is simply not sustainable in the long term.[28]

These net constraints are only a small part of the equation, however.

Many Americans may understand the phrase *environmental crisis* as re-ferring specifically to climate change, which has attracted the lion's share of environmental publicity in the last decade, but most scientists would join Buell in noting: "Even a short list of current environmental crises is of necessity quite long. At the least it must include an energy (and also other resources) crisis; a multifactorial waste crisis; . . . a wet-lands crisis; a food production crisis; a crop diversity crisis; a forest cri-sis; a soils crisis; an ocean crisis; a freshwater crisis; a biodiversity crisis; an acid rain crisis; an ozone hole crisis; a global warming crisis; an en-vironmental toxification crisis; a global disease crisis; a population cri-sis; and a growth or development crisis."[29] Most readers will be aware of at least one or two of these crises and possibly also aware that, despite the green rhetoric that is now standard in most advertising, very little has been done to address them—in fact, almost every one of them has become more grave since they were first brought to public attention.[30] As Sarah S. Amsler observed in 2010: "The spectre of crisis now casts an urgent but oddly bearable shadow on everyday life. It appears through documentaries on the science of climate change and video footage of melting ice: we manage it with recycling bins and reusable bags."[31]

Amsler was referring specifically to Al Gore's 2006 documentary *An Inconvenient Truth*, which represented a high-water mark of climate change awareness but offered only minor, mostly individualistic sug-gestions for action. As Michael Pollan noted in a review, "the really dark moment" of the film was not its visualization of the Atlantic Ocean washing over Manhattan—which most audience members were accustomed to seeing in blockbuster disaster movies such as *The Day After Tomorrow* (2004): "[That moment] came during the closing cred-its, when we are asked to . . . change our light bulbs. That's when it got really depressing. The immense disproportion between the magnitude of the problem Gore had described and the puniness of what he was asking us to do about it was enough to sink your heart."[32] The "puni-ness" of Gore's suggestions, mirrored in the baby-step solutions that mainstream environmental organizations and liberal politicians pro-posed throughout the period 2000–2010, reflected his acceptance of mainstream economic growthmanship. But it also represented a prag-matic awareness of America's lukewarm support for environmental ini-tiatives. According to a 2013 poll on Americans' policy priorities, "deal-ing with global warming" was the lowest priority of the twenty options presented, behind even "dealing with global trade" and "dealing with moral breakdown." The more general "protecting [the] environment" came in at number twelve, just behind "reforming [the] tax system."[33]

In this context—with a public that still implicitly expects an indefi-
nite continuation of business as usual enabled by a miraculous technological fix, even with an unprecedented environmental crisis on the
horizon—studying the rare individuals and groups that have actually
responded to environmental crisis in a proportionate way provides
one vision of what "taking crisis seriously" might look like.[34] Instead
of suppressing or ignoring the alarming information presented above,
we should use it as motivation for political action and the construction
of new political, philosophical, and affective orientations appropriate
to the era of climate change and "tough oil."[35] Peakists overestimated
the timing and immediate consequences of petroleum depletion and
underestimated the potential of unconventional fossil fuels, but their
sense of an imminent ecological crisis was certainly justified. They adopted and acted on an ecological identity that critiques and opposes
the normative but increasingly problematic dominant social paradigm
of unlimited abundance and technological solutionism described in
chapter 1. Furthermore, they were engaged in the kinds of behaviors
that we might hope all Americans would undertake to cut our carbon
emissions, necessary though not sufficient actions such as using public
transit, making their homes more energy efficient, driving and flying
less, bicycling more, and consuming responsibly. They argued, as one
believer put it, that environmental issues such as peak oil and climate
change are, at their core, "moral, ethical and spiritual" problems, as
opposed to "technical, scientific, engineering," or "economic" problems, and advised that "if we do not approach it from this direction
our chances for surviving, much less creating a more just, kind and humane society are very poor indeed."[36] This might have seemed a radical
claim only a decade ago, but that time has surely passed.

While the kinds of actions that peakists took had a negligible effect
on actual carbon emissions, the potential for energy conservation to
significantly reduce carbon emissions is often understated.[37] More importantly, such acts might have a tangible effect on the environmental
crisis of the will that enables the continuing carbon catastrophe—the
sense that Americans (and human beings in general) are not willing
or even able to change their lifestyles and make sacrifices on behalf
of the common good. As the sociologist Kari Marie Norgaard observed
from her ethnographic research into climate change quiescence in the
United States and Norway: "Individuals see that smart people around
them continue to carry on highly consumptive behaviors, even in the
face of knowledge about their consequences, and assume that others
are too self-interested to be motivated for change."[38] In this light, per-

sonal responses are important not only for their potential to reduce actual emissions but as an environmental communications strategy. They signal to friends, family, coworkers, neighbors, and complete strangers that, despite the imperative to consume, Americans do possess a concern for the commonweal and a willingness to change and even sacrifice.

Studying peak oil believers also provides a lesson about the causes of personal as well as political inaction. Sociologists and social psychologists such as Vladas Griskevicius, Niamh Murtagh, and Hal Herschfield have shown that environmental considerations are deeply dependent on social context, but the silencing of (currently) marginal environmental discourses has received less attention.[39] Almost uniformly, respondents to my 2013 questionnaire reported no longer even attempting to engage others in conversations about petroleum dependency and climate change. A New York man said: "For several years . . . I tried to educate everyone around me and encourage a 'coming together' to develop a local support system that I thought would be paramount as we commence with decline. However, there just isn't much of a 'listening' for the conversation." A man in his fifties observed that the response of friends and family was "not good": "It doesn't make for good cocktail party conversation . . . [and] caused some people to question my sanity."[40]

Of course, the social dynamic of marginalization that these peakists described is not limited to apocalyptic predictions about oil depletion. It extends to a number of potentially uncomfortable or unpleasant topics in the twenty-first-century United States, such as the roots of the recent recession, widening economic inequality, rates of incarceration that are seven times as high as those in similar industrialized nations, the ethics of drone strikes, the growing surveillance state, and climate change, to name just a few. Millions of conversations that never occur because one party is concerned about being perceived as a "Debby Downer" constitute the social production of an increasingly problematic normality. They are, in sum, exactly the kind of missed collective conversation that most observers believe the United States desperately needs, and their absence exerts a silent influence on the nation's ability to respond to the multitudinous challenges of our time. In our post–*Citizens United* era, it is all too easy to forget that, as Nina Eliasoph put it, "without the vibrant public sphere, democratic citizenship is impossible": "Only plain talk, between citizens, can knit the bonds necessary for a more humane society."[41]

To illustrate the impact of ignoring societal problems, a peakist from

Washington caricatured the "knee-jerk reaction" she often received when discussing resource depletion: there's "no evidence, you're a pessimist anyway, *yeah and my house is going to be under water next year as well.*"[42] As Hurricane Katrina, superstorm Sandy, and so many other recent weather events have shown, there is a high price to be paid for avoiding difficult topics, and it compounds over time. It would be comforting to maintain that peak oil and climate change are unrelated threats, but they are, in fact, two sides of the same oily coin. Both are global carbon-based problems that threaten disastrous consequences if we do not transition away from fossil fuels. If the Age of Oil ends soon, as peakists believed, it would, indeed, constitute a calamitous event; should we continue burning fossil fuels at the present rate for decades, the consequences of climate change will be even worse than current estimates predict.

Denial of environmental science has rightly received a great deal of attention in the United States in the last decade, primarily focusing on the conservative crusade to obfuscate and politicize climate science.[43] But there are other forms of denial as well. The social psychologists Susan Opotow and Leah Weiss have described the assignation of responsibility to "higher authorities or legitimate decision makers" as itself a form of denial of self-involvement.[44] And of growing importance as the consequences of our reliance on fossil fuels become more widely accepted, the sociologist Stanley Cohen has argued, is "implicatory denial," which admits the basic facts but minimizes "the psychological, political or moral implications that conventionally follow" from them, such as the ethical imperative to respond.[45] Sociologists such as Norgaard have demonstrated that these various forms of denial are socially and culturally constructed: different cultures and subcultures have differing norms of attention and emotion, meaning the kinds of information we pay attention to, ignore, and allow to influence us emotionally in our everyday lives. First, we rely on the "cognitive traditions" of our culture or subculture, which tell us "whether to pay attention to a given idea or event in a given moment or not." Then we rely on "emotion norms," which "prescribe the socially appropriate range, intensity, duration, and targets of feelings in different situations" and, thereby, "set the standard for what an individual 'ought' to feel in a given context." Norgaard notes that, in contrast to purely psychological theories of denial, "the notion of socially organized denial emphasizes that ignoring occurs in response to social circumstances and is carried out through a process of social interaction."[46] In my surveys, peakists described in great detail the kinds of subtle and not-so-subtle social pressure exerted by their friends, family, and coworkers and the cost of going against the grain.[47] Studying radical environmentalists such as peakists illuminates the contours of the social organization of denial.

Peak Politics

If peakists are to be respected for taking environmental crisis seriously, this book has been critical of the manner in which many of them responded. Here we find the convergence of the two primary issues of *Peak Politics*: concerns about environmental crisis and the influence of the libertarian shift. My respondents, like peak oil authors, were well aware that individualistic, survivalist responses to the problems of peak oil and climate change would not be sufficient, but many saw few alternatives. This has less to do with a personal failure of imagination and more to do with the historical moment. In a recent book on "catastrophism" and politics, Eddie Yuen argued that "waking up" to the current environmental situation "in the context of alienation is profoundly disempowering" and appropriately distinguished this contemporary experience from political consciousness raising during the 1960s, a decade of expanding democratic possibilities. "To understand" apathy about environmental challenges, Yuen asserted, we "must look at the conditions of atomization, depoliticization, powerlessness, and alienation that afflict the U.S. body politic generally."[48]

This is to say that changes in environmental policy and the human activities that have brought on various environmental crises must begin with a political transformation. As I argued in chapter 3, beginning in the 1960s libertarian activists and their wealthy backers slotted their then-radical political philosophy into the long tradition of American individualism and, thereby, effected a monumental transformation of American political culture. Capitalizing on the sense of inevitable growth and progress borne of oil-fueled post–World War II abundance, government became primarily a matter of optimizing markets for maximum efficiency. The market was not merely an answer to potentially divisive questions about the distribution of wealth and environmental deterioration but, as Daniel T. Rodgers observed, "stood for a way of thinking about society with a myriad of self-generated actions for its engine and optimization as its natural and spontaneous outcome."[49] The Internet that we use every day, developed and popularized in the heyday of market fundamentalism, reflected the dominant ideology of its era. Although we should hope that digital spaces will eventually deliver on their substantial democratic potential, we should recognize that they can also serve to atomize and depoliticize their inhabitants. During the recession of 2007–8, when the failure of market fundamentalism became too big to ignore, liberals finally began to push back against the spread of libertarian ideals. But by this point they were bucking the tide of three decades of gerrymandering[50] and conservative warnings about the evils of big government, praise for market efficiency, and the targeted evisceration of the public sphere.

With environmental issues ranking as a low priority, and in the absence of effective popular challenges to the status quo, the only politically acceptable means to respond to climate change and resource depletion is the favorite of neoliberals and libertarians alike: the market. The mainstream response to concerns about oil depletion, which became widespread in the United States in 2008, was, as Peter North summarized it, "that market economies are creative enough to solve problems through solutions of which we cannot as yet even dream, and that technology will overcome contemporary carbon crises."[51] This response was even echoed by dedicated scholars of petroleum.[52] For American politicians concerned with climate change, the only imaginable solution is a cap-and-trade system, despite common knowledge about the abuse of emissions markets overseas and the failure of similar environmental markets in recent US history.[53] This is a libertarian solution that is enthusiastically advocated by liberals, Democrats, Republicans, and conservatives alike. Although climate change has been described as "the greatest market failure the world has seen" since greenhouse gas emissions have remained external to the emitter's cost, the only solution is somehow to create even more markets.[54] But, we should ask ourselves, is this faith in market solutions and the concomitant opposition to regulation not, in part, a reflection of conservative intellectual influence? In this light, we might better understand the Nebraska peakist's claim: "The problem of peak oil, like overpopulation and nuclear proliferation, is about collective action and long-term planning. The problem is not a shortage of hydrocarbons [nor, one might add, an excess of hydrocarbons] but a shortage of wisdom."[55]

Wisdom, however, is one resource that is always renewable. This book is descriptive and analytic, and it is beyond its scope to offer prescriptions. However, I can conclude with the simple observation that the adoption of libertarian solutions is not a sound means to maintain the status quo, let alone achieve a future that is more equitable, more just, and more sustainable. Wisdom would include a recognition of the gradual trend toward libertarian and neoliberal ideals in political culture, not only in the United States, but also around the world. Wisdom would also require understanding that the solutions to global problems such as climate change and eventual resource depletion must be political as well as personal. With an awareness of our perilous moment in history, let us learn from our past missteps and collectively chart a better way forward.

Survey Data

	First survey	Second survey
Date conducted	January 2011	July 2011
Number of respondents	1,128	628

If you had to quantify your level of certainty in the fundamental theory of peak oil—that global oil production will peak in the next decade (if it hasn't already), and that this event will have grave and potentially apocalyptic effects on the United States and around the world—on a scale of 1 to 10, what would it be? 1 is disbelief, 10 is complete certainty.

1	.5	.7
2	.3	.7
3	.8	0
4	.3	.4
5	3.4	0
6	3.9	1.5
7	10.3	7.8
8	18.3	19.0
9	24.5	26.9
10	37.8	42.9

Which country/region are you in?		
United States	77.3	70.2
Canada	8.0	9.7
Central America	.3	.5
South America	.3	0
Western Europe	7.9	10.7
Eastern Europe	.9	1.6
Africa	.3	0
Asia	.8	1.6
Other	4.3	5.8

What is your age?		
0–20	.1	. . .
21–25	1.8	. . .
26–30	5.4	. . .
31–35.	8.7	. . .
36–40	9.0	. . .

(continued)

	First survey	Second survey
41–45	13.6	. . .
46–50	14.1	. . .
51–55	17.0	. . .
56–60	14.1	. . .
61–65	9.4	. . .
66–70	3.5	. . .
71–75	2.3	. . .
76–80	.4	. . .
81+	.4	. . .
What is your gender?		
Male	83.9	73.4
Female	15.5	22.8
Rather not answer	.7	2.6
If you are in the United States, do you identify as . . .		
White	. . .	88.8
Black/African American	. . .	1.1
Asian	. . .	0
Hispanic/Latino	. . .	1.9
American Indian	. . .	.7
Pacific Islander/Hawaiian	. . .	.7
Other	. . .	3.7
Rather not answer	. . .	4.1
If you are in the United States, what would you identify as your race or ethnicity?		
Caucasian	90.7	. . .
African American	.5	. . .
Latino	1.2	. . .
Asian American	.6	. . .
Native American	.8	. . .
Other	5.8	. . .
What is your religious preference?		
Protestant	. . .	5.5
Christian (nonspecific)	. . .	9.0
Catholic	. . .	5.5
Jewish	. . .	2.9
Mormon	. . .	.5
Other specific	. . .	11.6
None	. . .	50.9
Undesignated	. . .	14.0
What is the highest level of education you have completed?		
Did not graduate high school	.4	1.1
High school	6.0	7.1
College	44.2	46.1
Trade or technical school	6.2	7.1
Master's degree	26.7	24.7
Ph.D.	9.4	9.4
Other professional degrees	7.1	9.3
How would you describe your political views?		
Very conservative	. . .	4.3
Conservative	. . .	7.4
Moderate	. . .	29.9

	First survey	Second survey
Liberal	. . .	28.7
Very liberal	. . .	27.4
How do you identify yourself politically?		
Anarchist	6.5	9.0
Socialist	10.2	10.5
Progressive	20.1	20.1
Liberal	15.4	13.9
Independent	21.9	22.3
Moderate	10.2	8.4
Conservative	6.4	4.2
Libertarian	9.4	11.6
At the height of your interest in the subject, how often did you visit peak oil Web sites (approximately)?		
Many times a day	24.6	39.3
Once a day	42.3	42.3
Once a week	18.9	11.6
Once a month	6.4	2.2
Once a year	3.1	.7
Never	4.8	3.0
What initially sparked your interest in this topic?		
1970s oil crisis	14.6	. . .
The Gulf War	4.8	. . .
9/11	5.2	. . .
Wars in Iraq and Afghanistan	8.7	. . .
Hurricane Katrina	1.9	. . .
Gas prices	10.8	. . .
BP oil spill	1.3	. . .
A book	28.6	. . .
Friend/family/coworker	7.5	. . .
Have you found it difficult to speak to others about peak oil or "come out" as a peak oiler?		
Yes	29.3	. . .
Sometimes	36.7	. . .
Rarely	11.2	. . .
Not at all	21.4	. . .
Never tried	.3	. . .
Have you attended any in-person meetings with others interested in the same topic? If so, where?		
MeetUp groups	10.6	18.7
Conventions (regional or national)	9.9	11.2
Transition initiative	12.3	22.1
Lectures	27.9	31.2
No	61.4	53.0
What do you think the benefits of visiting sites like Peak Oil News & Message Boards (www.peakoil.com) have been for you? (Select all that apply.)		
Centralized location for information	72.7	. . .
Place to have questions answered	45.1	. . .
Good to be of service to newcomer	23.1	. . .
Comfort of knowing that other people have the same thoughts	54.9	. . .
Grow more of my own food	28.9	. . .
Invest differently	26.0	. . .
No benefits	.9	. . .
I don't visit peak oil Web sites	7.0	. . .

(continued)

	First survey	Second survey
How has your knowledge of peak oil affected your life? Please select all that apply.		
Changed occupation	12.8	. . .
Changed partner/spouse	2.5	. . .
Living in a Transition Town	3.3	. . .
Building a Transition Town	5.8	. . .
Living in other form of sustainable community	7.2	. . .
Building other form of sustainable community	15.6	. . .
Drive less	53.2	. . .
Purchased car with better MPG	32.3	. . .
Not affected my life	11.3	. . .
As a result of your knowledge of peak oil, have you done any of the following?		
Reduced energy usage at your current home	. . .	81.6
Moved to a smaller or more energy-efficient home	. . .	24.0
Prepared food or other supplies for yourself and your family	. . .	72.3
Changed your occupation	. . .	19.9
Moved to a Transition Town	. . .	2.6
Moved to some other form of sustainable community	. . .	12.4
Engaged in political activity related to peak oil	. . .	27.3
Please agree or disagree with the following statement: "Fictional portrayals of post-apocalyptic scenarios have influenced the vision that some 'peak oilers' have of what the post-peak period will be like."		
Strongly disagree	2.4	. . .
Disagree	14.7	. . .
Agree	68.5	. . .
Strongly agree	20.5	. . .
Please agree or disagree with the following statement: "Fictional portrayals of post-apocalyptic scenarios have influenced my image of what the post-peak period will be (and will not) be like."		
Strongly disagree	11.1	. . .
Disagree	27.1	. . .
Undecided	25.5	. . .
Agree	33.1	. . .
Strongly agree	3.1	. . .
Would you agree or disagree with the following statement: "We don't give everyone an equal chance in this country."		
Agree	. . .	82.2
Disagree	. . .	15.2
In recent years, has too much been made of the problems facing black people?		
Too much	. . .	10.1
Too little	. . .	34.5
Just right	. . .	12.7
Don't know	. . .	40.7
Do you think it should be legal or illegal for gay and lesbian couples to get married? How strongly do you feel about this question?		
Legal (strongly)	. . .	56.1
Legal (somewhat)	. . .	24.1
Illegal (somewhat)	. . .	3.3
Illegal (strongly)	. . .	3.3
No opinion	. . .	11.4
Which of the following websites have you checked regularly?		
Life After the Oil Crash	. . .	31.8
Peak Oil News	. . .	30.7

	First survey	Second survey
The Oil Drum	. . .	54.6
Energy Bulletin	. . .	39.4
Peak Oil Blues	. . .	15.5
Clusterfuck Nation	. . .	39.5
Which of the following books have you read?		
Richard Heinberg's *The Party's Over*	. . .	78.3
James Howard Kunstler's *The Long Emergency*	. . .	87.2
Matthew Simmons' *Twilight in the Desert*	. . .	77.1
James Howard Kunstler's *World Made by Hand*	. . .	78.1
Rob Hopkins' *Transition Handbook*	. . .	72.7
Dmitry Orlov's *Reinventing Collapse*	. . .	77.6
Which of the following do you see as most likely *for the country you currently reside in*? Please rate each on a scale of 0 to 10.		
Business as usual with no major consequences	. . .	1.9
Better technology and new energy sources allow for little change	. . .	2.7
Resource wars for remaining energy	. . .	7.8
A gradual shift into a lower-energy world ("powerdown")	. . .	5.7
A smaller population	. . .	7.1
A significant drop in quality of life	. . .	7.9
Apocalyptic scenario (i.e., violence, epidemics, die-off)	. . .	6.1
Which of the following do you see as most likely *in the less developed parts of the world*? Please rank each on a scale of 1 to 10.		
Business as usual with no major consequences	. . .	2.8
Better technology and new energy sources allow for little change	. . .	2.4
Resource wars for remaining energy	. . .	7.3
A gradual shift into a lower-energy world ("powerdown")	. . .	4.9
A smaller population	. . .	7.3
A significant drop in quality of life	. . .	6.3
Apocalyptic scenario (i.e., violence, epidemics, die-off)	. . .	6.8
Do most members of Congress deserve reelection?		
Yes	. . .	2.6
No	. . .	85.7
No opinion	. . .	11.2

Questionnaire, July 2013

How have your views about peak oil and energy depletion changed
since early 2011?

Has the exploitation of unconventional petroleum sources (such
as tar sands) and/or the utilization of new extraction methods
changed the way you see energy depletion and similar issues
playing out?

In terms of actions or preparations, what are you doing now, and
how is that different from what you were doing in 2011?

What was or is the impact of your beliefs in peak oil on your relationships with friends, family, co-workers, and/or loved ones?

Are you still as interested in the subject of energy depletion as you
were in 2011? If not, what kinds of similar concerns or interests, if any, do you now have?

Have you become involved in collective or political activities, such
as Transition Towns or political movements? If so, what is the
goal? If not, why?

What is your current occupation? What was it in 2011?

Notes

Unless otherwise noted, all URLs were current and active when this book went to press.

1. This estimate emerges from analysis of Web site membership and two surveys conducted for this book. For more details, see n. 51 below.
2. Although we typically think of identity as being determined by more traditional categories, such as race, class, gender, and sexual orientation, historians of environmentalism and conservatism can attest to the importance of self-chosen factors, such as dedicated conservative activism or deep ecology. See Mitchell Thomashow, *Ecological Identity: Becoming a Reflective Environmentalist* (Cambridge, MA: MIT Press, 1996); and Andrew Light, "What Is an Ecological Identity?" *Environmental Politics* 9, no. 4 (2000): 59–81.
3. "Porn. Peak Oil. Enjoy," http://www.youtube.com/watch?v=vAPf9V3_li0. As of this writing, the video has been viewed by over 300,000 people.
4. The values here reflect the differences between two surveys that I conducted in 2011. For information on survey responses and survey methodology, see n. 51 below.
5. Bryan Urstadt, "Imagine There's No Oil: Scenes from a Liberal Apocalypse," *Harper's*, August 2006, 31–40.
6. Richard Heinberg, *Peak Everything: Waking Up to a Century of Decline* (Gabriola Island, BC: New Society, 2007).
7. In response to the question, "As a result of your knowledge of peak oil, have you done any of the following," 72 percent had "prepared food or other supplies for yourself

and your family," 82 percent had "reduced energy usage at your current home," 24 had "moved to a smaller or more energy-efficient home," 27 had "engaged in political activity related to peak oil," and only 3 percent had "moved to a Transition Town." The Transition Initiative, begun in the United Kingdom in 2007, has received the most attention. Transition is a franchise model of intentional communities based on Rob Hopkins, *The Transition Handbook: From Oil Dependency to Local Resilience* (White River Junction, VT: Chelsea Green, 2008). For scholarship (mostly British) on Transition Towns, see Amanda Smith, "The Transition Town Network: A Review of Current Evolutions and Renaissance," *Social Movement Studies* 10, no. 1 (2011): 99–105; Ian Bailey, Rob Hopkins, and Geoff Wilson, "Some Things Old, Some Things New: The Spatial Representations and Politics of Change of the Peak Oil Relocalisation Movement," *Geoforum* 41, no. 4 (2010): 595–605; and Peter North, "Eco-Localisation as a Progressive Response to Peak Oil and Climate Change: A Sympathetic Critique," *Geoforum* 41, no. 4 (2010): 585–94.

8. Manuel de Landa, *A New Philosophy of Society: Assemblage Theory and Social Complexity* (New York: Continuum, 2006). See also Graeme Chesters and Ian Walsh, *Complexity and Social Movements: Multitudes at the Edge of Chaos* (New York: Routledge, 2006); and Nick Srnicek, "Assemblage Theory, Complexity and Contentious Politics: The Political Ontology of Gilles Deleuze" (master's thesis, University of Western Ontario, 2007).

9. In response to my survey question, "Have you been more or less engaged in the following activities since learning about peak oil? Please put on a scale from 1 to 5, with 1 for much less engaged, 3 for no change, and 5 for much less engaged," American peakists averaged 2.79 for the option "attending rallies, marches or protests." There are, of course, many varieties of political activity, of which participation in formal electoral politics and traditional political dissent are only the most public and widely acknowledged. Whether the retreat of some peakists from these types of political activities was accompanied by an increase in other activities that might well be termed *political* is an important question but one that is slightly beyond the methodology of this book.

10. "Oily Cassandra," personal communication, February 1, 2010.

11. "Porn. Peak Oil. Enjoy."

12. Respondent 9567344. Survey respondents are identified by their randomly generated response numbers.

13. Robert N. Bellah, Richard Madsen, William M. Sullivan, Ann Swidler, and Steven M. Tipton, *Habits of the Heart: Individualism and Commitment in American Life* (Berkeley and Los Angeles: University of California Press, 1985), 23.

14. Robert Putnam, *Bowling Alone: The Collapse and Revival of American Community* (New York: Simon & Schuster, 2000), 23.

15. For critiques, see Pamela Paxton, "Is Social Capital Declining in the United States? A Multiple Indicator Assessment," *American Journal of Sociol-*

ogy 105, no. 1 (1999): 88–127; and Everett Ladd, *The Ladd Report* (New York: Free Press, 1999). On the acceptance of the decline of social capital, see the symposium on *Bowling Alone* in *Contemporary Sociology* 30, no. 3 (2000): 223–30; and Stefano Bartolini, Ennio Bilancini, and Maurizio Pugno, "Did the Decline in Social Capital Decrease American Happiness? A Relational Explanation of the Happiness Paradox," *Social Indicators Research* 110, no. 3 (2013): 1033–59. Analyzing data from the World Values Survey, the sociologist Wayne Baker noticed a similar trend. See his *America's Crisis of Values: Reality and Perception* (Princeton, NJ: Princeton University Press, 2005), 56–60.

16. Eric Klinenberg, *Going Solo: The Extraordinary Rise and Surprising Appeal of Living Alone* (New York: Penguin, 2012), 4–5.
17. See, e.g., Sean Wilentz, *The Age of Reagan: A History, 1974–2008* (New York: Harper, 2008); Steven F. Hayward, *The Age of Reagan: The Conservative Counterrevolution, 1980–1989* (New York: Random House, 2009); and Jerome L. Himmselstein, *To the Right: The Transformation of American Conservatism* (Berkeley and Los Angeles: University of California Press, 1989), 13.
18. Frank Newport, "Americans Say Reagan Is the Greatest U.S. President," Gallup, February 18, 2011, http://www.gallup.com/poll/146183/americans -say-reagan-greatest-president.aspx. Nineteen percent of respondents named Reagan as the greatest president; Abraham Lincoln came in second, at 14 percent.
19. Ezra Klein, "Unpopular Mandate: Why Do Politicians Reverse Their Positions?" *New Yorker*, June 25, 2012, 30–33.
20. Kim Phillips-Fein, "Conservatism: A State of the Field," *Journal of American History* 98, no. 3 (2011): 723–43, 742.
21. Wilentz, *The Age of Reagan*, 8.
22. See Steven Teles and Daniel A. Kenney, "Spreading the Word: The Diffusion of American Conservatism in Europe and Beyond," in *Growing Apart? America and Europe in the Twenty-First Century*, ed. Jeffrey Kopstein and Sven Steinmo (New York: Cambridge University Press, 2008), 136–69.
23. *Giant*, George Stevens (1956); *There Will Be Blood*, Paul Thomas Anderson (2007).
24. For EROI data, see C. J. Cleveland, R. Costanza, C. A. S. Hall, and R. Kaufmann, "Energy and the U.S. Economy: A Biophysical Perspective," *Science* 225 (1984): 890–97; C. J. Cleveland, "Net Energy from Oil and Gas Extraction in the United States, 1954–1997," *Energy* 30, no. 5 (2005): 769–82; and C. A. S. Hall, C. J. Cleveland, and R. Kaufmann, *Energy and Resource Quality: The Ecology of the Economic Process* (New York: Wiley, 1986). For an overview of current EROI data for petroleum and other energy sources, see Ajay K. Gupta and Charles A. S. Hall, "A Review of the Past and Current State of EROI Data," *Sustainability* 3, no. 10 (2011): 1796–1809. For a discussion of the importance of EROI, see Charles A. S. Hall, "Editorial

Introduction to Special Issue on New Studies in EROI (Energy Return on Investment)," *Sustainability* 3, no. 10 (2011): 1773–77.

25. Charles A. S. Hall, "Unconventional Oil: Tar Sands and Shale Oil—EROI on the Web, Part 3 of 6," The Oil Drum, April 15, 2008, http://www .theoildrum.com/node/3839; Cutler J. Cleveland and Peter A. O'Connor, "Energy Return on Investment of Oil Shale," *Sustainability* 3 (2011): 2307–33. For fossil fuels, a lower EROI means increased carbon emissions, a related and increasingly important aspect of energy depletion.

26. See Praveen Ghanta, "Is Peak Oil Real? A List of Countries Past Peak," The Oil Drum, July 18, 2009, http://www.theoildrum.com/node/5576. Although there is a great deal of debate about whether a few major countries, such as Saudi Arabia, are at their peak, there is little debate about others, such as the United States, Iraq, and Mexico.

27. For example, while 120 giant oil fields (i.e., fields producing 100,000 barrels per day for more than one year) were discovered in the 1960s, fewer than 60 were discovered in the 1980s, fewer than 40 in the 1990s, and so on. Although giants make up only 1 percent of total world oil fields, they accounted for over 60 percent of world oil production as of 2005. See Mikael Höök, Robert Hirsch, and Kjell Aleklett, "Giant Oil Field Decline Rates and Their Influence on World Oil Production," *Energy Policy* 37, no. 6 (2009): 2262–72.

28. There are two potential motivations for nations to overestimate reserves. The first is to avoid a panic about running out of oil, which could raise prices in the short term (and, thus, prove profitable for net exporters) but also stimulate development of renewable sources. The second is that OPEC ties production to proven reserves. By overstating proven reserves, OPEC members would free themselves to extract more oil. For example, in the 1980s, a number of OPEC countries announced massive jumps in proven reserves almost overnight, with no explanation for the change.

29. Jeff Hatlen quoted in Jad Mouawad, "Oil Innovations Pump New Life into Old Wells," *New York Times*, March 5, 2007, A1.

30. Richard H. K. Vietor, *Energy Policy in America since 1945: A Study of Business-Government Relations* (New York: Cambridge University Press, 1984), 199. See also American Petroleum Institute, *Petroleum Facts and Figures* (New York, 1971), 101.

31. For an extrapolation of the date of peak oil from the Energy Information Administration 2011 report, see Eric L. Garza, "The US Energy Information Administration's Faulty Peak Oil Analysis," *Energy Bulletin*, August 19, 2011, http://www.energybulletin.net/stories/2011-08-19/us-energy -information-administrations-faulty-peak-oil-analysis.

32. Duncan Clark, "UK Will Face Peak Oil Crisis within Five Years, Report Warns," *Guardian*, October 29, 2008, http://www.theguardian.com/ environment/2008/oct/29/fossil-fuels-oil.

33. Nick A. Owen, Oliver R. Inderwildi, and David A. King, "The Status of

Conventional World Oil Reserves—Hype or Cause for Concern?" *Energy Policy* 38, no. 8 (2010): 4743–49.

34. "Porn. Peak Oil. Enjoy."
35. Though it has existed for quite some time, the phrase *resource wars* was popularized by Michael Klare's *Resource Wars: The New Landscape of Global Conflict* (New York: Henry Holt, 2002).
36. See, e.g., Dieter Helm, "Peak Oil and Energy Policy—a Critique," *Oxford Review of Economic Policy* 27, no. 1 (2011): 68–69; and Marian Radetzki, "Peak Oil and Other Threatening Peaks—Chimeras without Substance," *Energy Policy* 38, no. 11 (2010): 6566–69. Stephen M. Gorelick's *Oil Panic and the Global Crisis: Predictions and Myths* (Hoboken, NJ: Wiley-Blackwell, 2009) offers the most detailed scholarly refutation of the peak oil scenario.
37. E. D. Porter, "Are We Running Out of Oil?" Discussion Paper no. 081 (Washington, DC: American Petroleum Institute, 1995), 1.
38. Daniel Yergin, *The Prize: The Epic Quest for Oil, Money and Power* (New York: ABC, 1996), 194.
39. James Akins, "The Oil Crisis: This Time the Wolf Is Here," *Foreign Affairs* 51, no. 3 (1973): 462–90.
40. Shane Mulligan, "Energy, Environment, and Security: Critical Links in a Post-Peak World," *Global Environmental Politics* 10, no. 4 (2010): 79–100, 87.
41. This incompatibility, detailed in the conclusion, can be measured in a number of ways, such as ecological footprint, human appropriation of net primary productivity, and ecosystem services indices. See Global Footprint Network, *Global Footprint Network 2012 Annual Report*, 2012, http://www.footprintnetwork.org/images/article_uploads/2012_Annual_Report.pdf; Fridolin Krausmann, Karl-Heinz Erb, Simone Gingrich, Helmut Haberl, Alberte Bondeau, Veronika Gaube, Christian Lauk, Christoph Plutzar, and Timothy D. Searchinger, "Global Human Appropriation of Net Primary Production Doubled in the 20th Century," *Proceedings of the National Academy of Sciences* 110, no. 25 (2013): 10324–29; and WWF International, *Living Planet Report 2012: Biodiversity, Biocapacity and Better Choices*, 2012, http://awsassets.panda.org/downloads/1_lpr_2012_online_full_size_single_pages_final_120516.pdf. A useful summary can be found in James Gustave Speth, *The Bridge at the End of the World: Capitalism, the Environment, and Crossing from Crisis to Sustainability* (Ann Arbor, MI: Caravan, 2008), 17–45.
42. Tyler Priest, "The Dilemmas of Oil Empire," *Journal of American History* 99, no. 1 (2012): 236–51, 236.
43. Respondent 9569301.
44. On the oil industry, see Ida M. Tarbell, *The History of the Standard Oil Company* (New York: McClure, Phillips, 1904); Kendall Beaton, *Enterprise in Oil: A History of Shell in the United States* (New York: Appleton-Century-Crofts, 1957); Harold F. Williamson, Ralph L. Andreano, and Arnold R. Daum, *The American Petroleum Industry: The Age of Energy, 1899–1959* (Evanston,

IL: Northwestern University Press, 1963); and Anthony Sampson, *The Seven Sisters: The Great Oil Companies and the World They Shaped* (New York: Bantam, 1975). On oil politics, see Roger M. Olien and Diana Davids Olien, *Oil and Ideology: The Cultural Creation of the American Petroleum Industry* (Chapel Hill: University of North Carolina Press, 2000); F. William Engdahl, *A Century of War: Anglo-American Oil Politics and the New World Order* (Ann Arbor, MI: Pluto, 2004); and Yergin, *The Prize*. On wildcatters, see Edgar Wesley Owen, *Trek of the Oil Finders: A History of Exploration for Petroleum* (Tulsa, OK: American Association of Petroleum Geologists, 1975); and Lon Tinkle, *Mr. De: A Biography of Everette Lee DeGolyer* (New York: Little, Brown, 1970). On production in individual states or regions, see Paul Sabin, *Crude Politics: The California Oil Market, 1900–1940* (Berkeley and Los Angeles: University of California Press, 2004). On oil spills, see Robert Olney Easton, *Black Tide: The Santa Barbara Oil Spill and Its Consequences* (New York: Delacorte, 1972); and Noah Berlatsky, ed., *The Exxon Valdez Oil Spill* (Farmington Hill, MI: Greenhaven, 2011). Two notable exceptions to these categories include Maurice Berkeley Green, *Eating Oil: Energy Use in Food Production* (Boulder, CO: Westview, 1978); and Peter H. Spitz, *Petrochemicals: The Rise of an Industry* (New York: Wiley, 1988).

45. For recent popular works, see, among many others, Peter Maass, *Crude World: The Violent Twilight of Oil* (New York: Knopf, 2009); Lisa Margonelli, *Oil on the Brain: Adventures from the Pump to the Pipeline* (New York: Broadway, 2008); and Paul Roberts, *The End of Oil: On the Edge of a Perilous New World* (Boston: Houghton Mifflin, 2004). Notable scholarly publications include Timothy Mitchell, *Carbon Democracy: Political Power in the Age of Oil* (New York: Verso, 2011); Matthew T. Huber, *Lifeblood: Oil, Freedom, and the Forces of Capital* (Minneapolis: University of Minnesota Press, 2013); Brian C. Black, *Crude Reality: Petroleum in World History* (New York: Rowman & Littlefield, 2012); Stephanie LeMenager, *Living Oil: Petroleum Culture in the American Century* (New York: Oxford University Press, 2014); Ross Barrett and Daniel Worden, eds., *Oil Culture* (Minneapolis: University of Minnesota Press, 2014); Imre Szeman, Patricia Yaeger, and Jennifer Wenzel, eds., *Fueling Culture: Energy, History, Politics* (New York: Fordham University Press, 2015); and many others mentioned in these pages.

46. In 2013, Americans consumed an average of 18.89 million barrels of petroleum per day ("How Much Oil Is Consumed in the United States?" US Energy Information Administration, n.d., http://www.eia.gov/tools/faqs/faq.cfm?id=33&t=6). Since many products that Americans purchase are produced overseas and require oil for shipping and petrochemicals for manufacture, the actual figure is higher.

47. "Combien de pétrole dans mon ordinateur?" Actualité, n.d., http://www.linternaute.com/actualite/savoir/07/petrole-yaourt/7.shtml.

48. Jay Hakes, "Introduction: A Decidedly Valuable and Dangerous Fuel," *Journal of American History* 99, no. 1 (2012): 19–23, 19. See also US Energy

Information Administration, "U.S. Refinery Yield," n.d., http://www.eia
.gov/dnav/pet/PET_PNP_PCT_DC_NUS_PCT_A.htm.

49. Carl Solberg, *Oil Power* (New York: Signet, 1976), 8.
50. They have critiqued the immediacy of the threat (Helm, "Peak Oil and Energy Policy"; Radetzki, "Peak Oil and Other Threatening Peaks"), discussed
the dangers of peak oil in relation to climate change (Peñuelas, Joseph, and
Jofre Carnicer, "Climate Change and Peak Oil: The Urgent Need for a Transition to a Non-Carbon-Emitting Society," *Ambio* 39 [2010]: 85–90), viewed
peak oil in the context of capitalist "energy imperialism" (John Bellamy
Foster, "Peak Oil and Energy Imperialism," *Monthly Review* 60, no. 3 [2008]:
12–33), and examined the founder of the peak oil theory, the former Shell
geologist M. King Hubbert (Emma Hemmingsen, "At the Base of Hubbert's
Peak: Grounding the Debate on Petroleum Scarcity," *Geoforum* 41, no. 4
[2010]: 531–40). Others have speculated how the United States (Jessica G.
Lambert and Gail P. Lambert, "Predicting the Psychological Response of the
American People to Oil Depletion and Declining Energy Return on Investment [EROI]," *Sustainability* 3, no. 11 [2011]: 2129–56) and the entire world
(Jörg Friedrichs, "Global Energy Crunch: How Different Parts of the World
Would React to a Peak Oil Scenario," *Energy Policy* 38, no. 8 [2010]: 4562–69)
will collectively respond to oil depletion and attempted to assess its poten-
tial impact on future tourism (Susan Becken, "Developing Indicators for
Managing Tourism in the Face of Peak Oil," *Tourism Management* 29, no. 4
[2008]: 695–705; James Leigh, "New Tourism in a New Society Arises from
'Peak Oil,'" *Tourismos: An International Multidisciplinary Journal of Tourism* 6,
no. 1 [2011]: 165–91), urban areas (Peter Newman, "Beyond Peak Oil: Will
Our Cities Collapse?" *Journal of Urban Technology* 14, no. 2 [2007]: 15–30),
public health (Nikhil Kaza, Gerrit-Jan Knaap, Isolde Knaap, and Rebecca
Lewis, "Peak Oil, Urban Form, and Public Health: Exploring the Connections," *American Journal of Public Health* 101, no. 9 [2011]: 1598–1606; P. Hanlon and G. McCartney, "Peak Oil: Will It Be Public Health's Greatest Challenge?" *Public Health* 122, no. 7 [2008]: 647–52; Brian S. Schwartz, Cindy L.
Parker, Jeremy Hess, and Howard Frumkin, "Public Health and Medicine in
an Age of Energy Scarcity: The Case of Petroleum," *American Journal of Public
Health* 101, no. 9 [2011]: 1560–67; Peter Winch, and Rebecca Stepnitz, "Peak
Oil and Health in Low- and Middle-Income Countries: Impacts and Potential Responses," *American Journal of Public Health* 101, no. 9 [2011]: 1225–34),
children's welfare (Paul Tranter and Scott Sharpe, "Escaping Monstropolis:
Child-Friendly Cities, Peak Oil and *Monsters, Inc.*," *Children's Geographies*
6, no. 3 [2008]: 295–308), globalization (Fred Curtis, "Peak Globalization:
Climate Change, Oil Depletion and Global Trade," *Ecological Economics*
69, no. 2 [2009]: 427–34), and transportation (Montira Watcharasukarn,
Shannon Page, and Susan Krumdieck, "Virtual Reality Simulation Game
Approach to Investigate Transport Adaptive Capacity for Peak Oil Plan-
ning," *Transportation Research* 46, no. 2 [2012]: 348–67). A few others have

even discussed responses to peak oil by individuals and groups by chart-
ing the psychology of "peak oil acceptance" (Lyle K. Grant, "Peak Oil as a
Behavioral Problem," *Behavior and Social Issues* 16, no. 1 [2007]: 65–88; Ugo
Bardi, "Peak Oil: Four Stages of a New Idea," *Energy* 34, no. 3 (2009): 323–26;
Adrian Atkinson, "Progress in Acknowledging and Confronting Climate
Change and 'Peak Oil,'" *City: Analysis of Urban Trends, Culture, Theory, Policy,
Action* 14, no. 3 [2010]: 314–22; Lin Zhao, Lianyong Feng, and Charles A. S.
Hall, "Is Peakoilism Coming?" *Energy Policy* 37, no. 6 [2009]: 2136–38), by
analyzing different kinds of peak oil narratives (John C. Pruit, "Considering
Eventfulness as an Explanation for Locally Mediated Peak Oil Narratives,"
Qualitative Inquiry 17, no. 2 [2011]: 197–203), and by examining the Transi-
tion Town movement in the United Kingdom (Smith, "The Transition Town
Network"; Bailey, Hopkins, and Wilson, "Some Things Old, Some Things
New"; North, "Eco-Localisation as a Progressive Response").

51. The first survey, conducted in January 2011, had 1,128 respondents; the
second, conducted in July 2011, had 628. Online surveys are especially
useful for research on hidden populations and populations with unpopular political views. See Kevin B. Wright, "Researching Internet-Based
Populations: Advantages and Disadvantages of Online Survey Research,
Online Questionnaire Authoring Software Packages, and Web Survey
Services," *Journal of Computer-Mediated Communication* 10, no. 3 (2005),
http://onlinelibrary.wiley.com/journal/10.1111/(ISSN)1083-6101. As a
survey method, snowball or chain sampling was chosen because it aids in
recruiting hidden populations. I allowed any interested party to complete
the surveys and asked respondents to forward the links to their peak oil
aware friends and acquaintances. The potential biases inherent in this
methodology—such as the accuracy of self-reported information and self-
selection bias—were counterbalanced by other research methodologies.
For example, the potential that the population of peakists who would answer an online survey (or even use a computer) might answer in systematically different ways than peakists who would not answer an online survey
(or use a computer) was disproved by interviews with believers at peak oil
conferences and Transition Towns. In response to the question, "If you
had to quantify your level of certainty in the fundamental theory of peak
oil—that global oil production will peak in the next decade (if it hasn't
already), and that this event will have grave and potentially apocalyptic
effects on the United States and around the world—on a scale of 1 to 10,
what would it be? 1 is disbelief, 10 is complete certainty," 93.8 percent
of respondents to the first survey and 85.6 percent of respondents to the
second survey answered 7 or more. The mean for each question was 8.70
and 8.85, respectively.

52. James Howard Kunstler, *The Long Emergency: Surviving the End of Oil, Climate Change, and Other Converging Catastrophes of the Twenty-First Century*
(New York: Atlantic Monthly Press, 2005).

53. Kathy McMahon, *"I Can't Believe You Actually Think That!" A Couple's Guide to Finding Common Ground about Peak Oil, Climate Catastrophe, and Economic Hard Times* (self-published, 2011).

54. Kathy McMahon, "Do You Have a Panglossian Disorder? or, Economic and Planetary Collapse: Is it a Therapeutic Issue?" Peak Oil Blues, November 13, 2007, http://www.peakoilblues.org/blog/?p=132.

55. See Barbara Ehrenreich, *Bright-Sided: How Positive Thinking Is Undermining America* (New York: Metropolitan, 2009); and Karen A. Cerulo, *Never Saw It Coming: Cultural Challenges to Envisioning the Worst* (Chicago: University of Chicago Press, 2006).

CHAPTER ONE

1. As this experience of conversion suggests, belief in peak oil bears at least a family resemblance to belief in religion proper. After becoming peak oil aware, adherents share a belief in worldly destruction and a new conception of proper or ethical conduct, practice rituals of prediction, prophecy, and preparation, and attempt to evangelize and convert their friends and family. See Bron Taylor, "Exploring Religion, Nature and Culture," *Journal for the Study of Religion, Nature and Culture* 1, no. 1 (2007): 5–24. Fifty percent of American respondents to my July 2011 survey said that their religious preference was "none," and 62 percent said that religion was "not very important" to them. In a similarly worded Gallup poll of all Americans taken a year before, only 14 percent of Americans answered "none," while 20 percent answered "not very important" (http://www.gallup.com/poll/1690/religion.aspx). That peakists are much less likely than other Americans to believe in God and describe religion as important to them suggests that, for some, peak oil is not compatible with or serves functions similar to those of traditional religions. For a minority, however, peak oil does imply a spiritual element. See, e.g., Carolyn Baker, *Sacred Demise: Walking the Path of Industrial Civilization's Collapse* (Bloomington, IN: iUniverse, 2009). Although peakism lacks a conception of the sacred or supernatural, belief in the peak oil theory certainly has religious dimensions.

2. Zhao, Feng, and Hall, "Is Peakoilism Coming?" Kurt Cobb has asked what members of the peak oil movement should call themselves. He considers *peakist, doomer, peak oil advocate, peak oiler, peak oil believer,* and *peak oil activist.* Kurt Cobb, "What Should Members of the Peak Oil Movement Call Themselves?" Energy Bulletin, December 2, 2007, http://www.energybulletin.net/node/38052.

3. Eric Foner, *Free Soil, Free Labor, Free Men: The Ideology of the Republican Party Before the Civil War* (New York: Oxford University Press, 1970), 4. Throughout this book, I use *ideology* and *belief system* interchangeably.

4. Respondent 9686672.

NOTES TO PAGES 18-21

5. Edward Shils, "Ideology: The Concept and Function of Ideology," in *The Encyclopedia of Philosophy* (8 vols.), ed. Paul Edwards (New York: Macmillan, 1967), 4:974–78.

6. The degree of belief in the theory of peak oil—the question of who is and who is not a peak oil believer—was most directly measured by responses to the question, "If you had to quantify your level of certainty in the fundamental theory of peak oil—that global oil production will peak in the next decade (if it hasn't already), and that this event will have grave and potentially apocalyptic effects on the United States and around the world—on a scale of 1 to 10, what would it be? 1 is disbelief, 10 is complete certainty." The mean for both surveys was 8.7, with over 95 percent answering 7 or higher.

7. Shils, "Ideology," 976.

8. See, e.g., Julian Darley, *High Noon for Natural Gas* (White River Junction, VT: Chelsea Green, 2004); Kenneth S. Deffeyes, *Beyond Oil: The View from Hubbert's Peak* (New York: Hill & Wang, 2005); Roberts, *The End of Oil*; Richard Heinberg, *The Party's Over: Oil, War, and the Fate of Industrial Societies* (Gabriola Island, BC: New Society, 2003); and Klare, *Resource Wars*. For energy use per capita per annum, see "Energy Use (Kg of Oil Equivalent per Capita)," World Bank, n.d., http://data.worldbank.org/indicator/EG.USE.PCAP.KG.OE. As Timothy Mitchell has noted, expanding demand and shrinking supply now mean that a new equivalent of Saudi Arabia must be discovered approximately every three years. Mitchell, *Carbon Democracy*, 232.

9. Heinberg, *Peak Everything*.

10. Respondents 9683173, 9590211, 9574456.

11. Although a religious awakening is one obvious parallel to this experience, political transformations also provide some interesting similarities. Many second-wave feminists, e.g., describe the "click" of a personal revelation about gender and patriarchy. See J. Courtney Sullivan and Courtney E. Martin, eds., *Click: When We Knew We Were Feminists* (New York: Seal, 2010).

12. Respondents 9576345, 9649893, 9688522, 9575766.

13. The term *doomer* originated in critiques of 1970s environmentalists but was embraced by some peakists. For more information about Malthusian environmentalists of that era and their cornucopian critics, see chapter 2.

14. Joseph Tainter, *The Collapse of Complex Societies* (Cambridge: Cambridge University Press, 1988); Jared Diamond, *Collapse: How Societies Choose to Fail or Succeed* (New York: Viking, 2005).

15. In response to the question, "Which of the following do you see as most likely for the country you currently reside in? Please rate each on a scale of 0 to 10," American peakists provided the following responses: business as usual with no major consequences = 2.01; better technology and new energy sources allow for little change = 2.26; a gradual shift into a lower-energy world = 5.71; apocalyptic scenario (i.e., violence, epidemics, die-off) = 6.37; a smaller population = 7.35; Resource wars for remaining

energy = 8.18; a significant drop in quality of life = 8.24. In discussions about energy futures, "business as usual" denotes a continuation of current energy regime and global capitalism without a major shift in course.

16. Respondents 9568924, 9569376, 9592049.
17. Respondent 9682014.
18. "Flycreeper," "Preparation Philosophy," "Life After the Oil Crash," http://www.lifeaftertheoilcrash.net/902385028 (now defunct).
19. Respondents 9574456, 9575542, 9565014, 9680012.
20. Respondents 9682014, 9573763, 9593771.
21. Seventy-two percent of American peakists answered yes to the question, "As a result of your knowledge of peak oil, have you done any of the following—Prepared food or other supplies for yourself and your family."
22. While this number might be considered evidence of politicization for some groups of Americans, many respondents already had long histories of political activism.
23. Respondent 9590253.
24. Sharon Astyk, The Chatelaine's Keys: Finding the Keys to the Future, http://sharonastyk.com (now defunct); Homesteading Today: Neighborly Help and Friendly Advice, http://www.homesteadingtoday.com; The Automatic Earth, http://theautomaticearth.com/Front-Page.html.
25. Respondent 9569860.
26. Respondent 9691422.
27. Whether these peakists' relationships actually ended because their partner could not agree with their new beliefs cannot be known, but what is significant is that these respondents identified their beliefs about energy scarcity as influencing the most intimate aspects of their personal lives.
28. Respondents 9566752, 9565782, 9560308, 9566339, 9569635, 9569898.
29. For example, the average American drove 7.6 percent fewer miles in 2013 than in 2004. While part of this shift can be explained by the increase in the proportion of workers who telecommute, many mid-sized cities have seen a renaissance in public (nonautomotive) transportation. See US PIRG Education Fund, *Transportation in Transition*, December 2013, http://www.uspirg.org/sites/pirg/files/reports/US_Transp_trans_scrn.pdf; and Joe Kloc, "America's Invisible Trolley System," *Newsweek*, June 5, 2014, http://www.newsweek.com/2014/06/13/americas-invisible-trolley-system-253455.html.
30. Respondent 9573358.
31. Respondent 9575122. An "upstream" investment relates to "earnings or operations at firms that are near or at the initial stages of producing a good or service": "For example, exploration and production are upstream operations for a large integrated oil company." See David L. Scott, *Wall Street Words: An A to Z Guide to Investment Terms for Today's Investor* (Boston: Houghton Mifflin, 2010), 8.
32. Stephen Leeb and Donna Leeb, *The Oil Factor: Protect Yourself and Profit from the Coming Energy Crisis* (New York: Warner Business, 2005); Stephen

Leeb and Glen Strathy, *The Coming Economic Collapse: How You Can Thrive When Oil Costs $200 a Barrel* (New York: Business Plus, 2006).

33. Respondents 9690636, 9697749.

34. For example, respondent 9686313.

35. Respondent 9680829.

36. The loudest call for divestment has come from the group 350.org. As McKenzie Funk showed in *Windfall: The Booming Business of Global Warming* (New York: Penguin, 2014), however, peakists were hardly alone in seeing potential profit in recent and future environmental transformations.

37. Respondents 9590211, 9569736. This is another area where peakists and conservative survivalists cross paths, of course, although few peakists seem to follow the latter's conspiracy theories about the Federal Reserve.

38. Respondent 9575122.

39. Respondents 9559141, 9564383, 9566499, 9568832, 9570009.

40. See Nancy B. McCaffrey, Sara Jane, and Douglas H. Hill, "The Localism Movement: Shared and Emergent Values," *Journal of Environmental Sustainability* 2, no. 2 (2012): 45–57; and Kim Humphrey, *Excess: Anti-Consumerism in the West* (London: Polity, 2009).

41. Respondents 9568679, 9567303, 9693509.

42. Respondents 9573763, 9567795, 9707269, 9568832, 9590310.

43. Joshua Piven, *The Complete Worst-Case Scenario Survival Handbook* (San Francisco: Chronicle, 2007); Max Brooks, *The Zombie Survival Guide: Complete Protection from the Living Dead* (New York: Broadway, 2003).

44. See Daniel Pargman, "The Great Reskilling," Resilience, January 18, 2010, http://www.resilience.org/stories/2010-01-18/great-reskilling.

45. Respondents 9562083, 9570857, 9602470, 9594065, 9576587.

46. See Michael Pollan, "The Food Movement, Rising," *New York Review of Books*, June 10, 2010, 31–33; Doreen Jakob, "Crafting Your Way out of the Recession? New Craft Entrepreneurs and the Global Economic Downturn," *Cambridge Journal of Regions, Economy and Society* 6, no. 1 (2013): 127–40; and David Brodwin, "Americans Leading a 'Do It Yourself Economy' as Washington Stalls," *U.S. News and World Report*, August 2, 2012, http://www.usnews.com/opinion/blogs/economic-intelligence/2012/08/02/americans-leading-a-do-it-yourself-economy-as-washington-stalls.

47. Respondents 9592316, 9590687, 9684005.

48. Respondents 9568487, 9590578, 9591555, 9563196. The Virginia man was one of only a hanful of (self-identified) peakists of color to take my surveys. In chapter 5, I discuss the racial dynamics of peakism in some detail.

49. Respondents 9564499, 9590983, 9562254.

50. See, e.g., Eduardo Porter, "For Insurers, No Doubts on Climate Change," *New York Times*, May 15, 2013, B1; and Funk, *Windfall*.

51. US Energy Information Administration, "Gasoline and Diesel Fuel Update," n.d., http://www.eia.gov/petroleum/gasdiesel.

52. Respondents 9563747, 9548945, 9589676, 9686974.

53. Respondent 9566200.
54. Respondents 9684150, 9584310. While we now tend to associate stockpil-
    ing supplies with right-wing survivalists, the practice has a longer history
    in the United States—witness the bomb shelters constructed during the
    Cold War. Most recent scholarship on the subject identifies the primary
    benefits of such actions as psychological, something that many peakists
    would not deny. Individualistic stockpiling is to be differentiated from the
    more collective tradition of national preparedness.
55. Respondents 9590181, 9695655, 9569428.
56. Respondents 9565782, 9576666, 9559965, 9569854, 9567071, 9558720,
    9561951.
57. Hanlon and McCartney, "Peak Oil"; Becken, "Developing Indicators for
    Managing Tourism"; Newman, "Beyond Peak Oil."
58. Brian Wilson, "Ethnography, the Internet, and Youth Culture: Strategies
    for Examining Social Resistance and 'Online-Offline' Relationships,"
    Canadian Journal of Education 29, no. 1 (2006): 307–28; Paul Tranter and
    Scott Sharpe, "Children and Peak Oil: An Opportunity in Crisis," *Interna-
    tional Journal of Children's Rights* 15, no. 1 (2007): 181–97.
59. Marc A. Smith, "Invisible Crowds in Cyberspace: Mapping the Social
    Structure of the Usenet," in *Communities in Cyberspace*, ed. Marc A. Smith
    and Peter Kollock (New York: Routledge, 1999), 195–218.
60. James Howard Kunstler, *The Geography of Nowhere: The Rise and Decline of
    America's Man-Made Landscape* (New York: Free Press, 1993); Kenneth T.
    Jackson, *Crabgrass Frontier: The Suburbanization of the United States* (New
    York: Oxford University Press, 1985); Jane Jacobs, *The Death and Life of
    Great American Cities* (New York: Random House, 1961).
61. Jonathan Yardley, "All Built out of Ticky-Tacky," *Washington Post*, May 30,
    1993, X03.
62. Scott Carlson, "A Social Critic Warns of Upheavals to Come," *Chronicle of
    Higher Education*, October 20, 2006, http://chronicle.com/article/A-Social
    -Critic-Warns-of/15941.
63. Similar works include Roberts, *The End of Oil*; Colin Campbell, *The Coming
    Oil Crisis* (Brentwood: Multi-Science, 2004); and David Goodstein, *Out of
    Gas: The End of the Age of Oil* (New York: Norton, 2005).
64. Hal Lindsay, *The Late, Great Planet Earth* (New York: Zondervan, 1970);
    "The Long Emergency . . . James Howard Kunstler," Peak Oil News, n.d.,
    http://peakoil.com/forums/post99674.html.
65. Kunstler, *The Long Emergency*, 254, 99, 239, 303, 274, 238.
66. Mark Hertsgaard, "Oil Supply Facts, Forecasts Provide Fuel for Thought,"
    Fort Wayne (IN) Journal-Gazette, September 11, 2005, 17.
67. Ben McGrath, "The Dystopians: Bad Times Are Boom Times for Some,"
    New Yorker, January 26, 2009, 40–49.
68. Dan Mitchel, "The World According to Kunstler," *New York Times*, July 30,
    2005, C5.

69. This means that approximately 1 of every 150–200 Americans visited Kunstler's site in 2008 alone.

70. Data provided by "StatCounter" via a link on Kunstler.com, http://beta .statcounter.com/p1456112/summary.

71. "Start here | Energy Bulletin," http://www.energybulletin.net/start.

72. Kris Can, "Public Service Message," http://www.kriscan.com/2008/04/16/ public-service-message (video has been removed).

73. ASPO International, http://www.peakoil.net.

74. Post Carbon Institute, http://www.postcarbon.org; interviews with Asher Miller and Brian, Denver, CO, October 12, 2009.

75. Interview with Bill Fickas, Denver, CO, October 11, 2009.

76. See, e.g., Naomi Oreskes and Erik M. Conway, "The Collapse of Western Civilization: A View from the Future," *Daedalus* 142, no. 1 (2013): 40–48; Diamond, *Collapse*; and Safa Motesharrei, Jorge Rivas, and Eugenia Kalnay, "Human and Nature Dynamics (HANDY): Modeling Inequality and Use of Resources in the Collapse or Sustainability of Societies," *Ecological Economics* 101 (2014): 90–102.

77. See, e.g., the "Peak Oil Gifts" page on Café Press.com, http://www .cafepress.com/+peak-oil+gifts.

78. "Oil Production: Is the Sky Falling?" Aspect Energy, LLC, 1775 Sherman St., Denver CO 80203.

79. Michael Lynch, "'Peak Oil' Is a Waste of Energy," *New York Times*, August 24, 2009, A21.

80. Kathy McMahon, "'Being Help' and 'Needing Help' (Revised)," Peak Oil Blues, May 12, 2006, http://www.peakoilblues.org/blog/2006/05/12/hello -world.

81. The Feisty Life, http://www.thefeistylife.com (now defunct; this venture had been changed to Berkshires Couples Counseling, http://www .berkshirecouplescounseling.com, but that too is now defunct); McMahon, *"I Can't Believe You Actually Think That!"*

82. "The Peak Oil Blues," Powering Down: A Journey of Preparation, July 26, 2006, http://poweringdown.blogspot.com/2006/07/peak-oil-blues.html; "Peak Oil Links," Dark Optimism: A Better Future for a Troubled World, n.d., http://www.darkoptimism.org/links.html; "Peak Oil Blues," Failing Gracefully: Finding Resilience in Uncertain Times, n.d., http:// failinggracefully.com.

83. "Episode #6: Peak Oil Blues," The Extraenvironmentalist, November 11, 2010, http://www.extraenvironmentalist.com/episode-6-peak-oil-blues; "Reality Report: Peak Oil Blues," "Global Public Media: Public Service Broadcasting for a Post Carbon World," December 18, 2006, http://old .globalpublicmedia.com (postings have since been removed; site is no longer active); "Peak Oil vs. 'Pathological Optimism,'" Ecoshock Radio, October 27, 2010, http://www.ecoshock.info/2010/10/peak-oil-vs-pathological-optimism

.html; "Gasoline Gangsters Episode 6: Women and Peak Oil," YouTube, February 19, 2011, http://www.youtube.com/watch?v=FQcK-5lE_SE.

84. Kathy McMahon, "Peak Oil Dating: What Is an 'Ideal PO Mate'?" Peak Oil Blues, August 5, 2008, http://www.peakoilblues.org/blog/?tag=ideal-peak -oil-mate.

85. From March to October 2011, e.g., there were over thirty-two thousand unique visitors. Peak Oil Blues, http://www.peakoilblues.org.

86. Kathy McMahon, "Grim Newlywed Sees Scary Future for Those He Loves," Peak Oil Blues, January 14, 2010, http://www.peakoilblues.org/blog/?p=1679.

87. "Praise," Peak Oil Blues, n.d., http://www.peakoilblues.org/blog/hey-what -do-people-say-about-dr-k.

88. Kathy McMahon, personal communication, November 18, 2009.

89. McMahon, "'Being Help' and 'Needing Help.'"

90. Direct quotations from Peak Oil Shrink readers. Kathy McMahon, "Stages of Peak Oil Awareness," Peak Oil Blues, September 22, 2010, http://www .peakoilblues.org/blog/?p=2381.

91. Kathy McMahon, personal communication, November 18, 2009.

92. Ecopsychology—a combination of psychology and psychotherapy that addresses the physical and psychological influence of environmental degradation—is a growing field. See, e.g., Andy Fisher, *Radical Ecopsychology: Psychology in the Service of Life* (Albany: State University of New York Press, 2002); and Theodore Roszak, Mary E. Gomes, and Allen D. Kanner, eds., *Ecopsychology: Restoring the Earth, Healing the Mind* (San Francisco: Sierra Club, 1995).

93. McMahon, "Do You Have a Panglossian Disorder?" One of the Peak Shrink's readers even turned the neologism into a song. "Panglossian," performed "to the tune of *Alone Again* by Henry Philips," was part of Michael Nabert's "Sustainability the Musical!" which was performed at the 2011 Hamilton (Ontario) Fringe Festival and is available online as an audio file. See "Sustainability the Musical," http://www.yourgreenfeat .com/Green_Feat/the_Musical.html.

94. On environmental toxification, see Theo Colborn, Dianne Dumanoski, and John Peter Meyes, *Our Stolen Future: Are We Threatening Our Fertility, Intelligence, and Survival?* (New York: Plume, 1997). On ocean acidification, deforestation, and mass extinction, see Elizabeth Kolbert, *The Sixth Extinction: An Unnatural History* (New York: Henry Holt, 2014).

95. Respondent 9592026.

96. Respondents 9574456, 9566499, 9591372, 9562162.

97. C. Vann Woodward, "The Age of Reinterpretation," *American Historical Review* 66, no. 1 (1960): 1–19, 3.

98. Neil D. Weinstein, "Optimistic Biases about Personal Risks," *Science* 246, no. 4935 (1989): 1232–33. See also Ahmed Abdel-Khalek and David Lester, "Optimism and Pessimism in Kuwaiti and American College Students,"

International Journal of Social Psychiatry 52, no. 2 (2006): 110–26; Edward C. Chang, Kiyoshi Asakawa, and Lawrence J. Sanna, "Cultural Variations in Optimistic and Pessimistic Bias: Do Easterners Really Expect the Worst and Westerners Really Expect the Best When Predicting Future Life Events?" *Journal of Personality and Social Psychology* 81, no. 3 (2001): 476–91; and Yueh-Ting Lee and Martin E. P. Seligman, "Are Americans More Optimistic Than the Chinese?" *Personality and Social Psychology Bulletin* 23 (1997): 32–40.

99. Ehrenreich, *Bright-Sided*, 46. *Motivation industry* is Ehrenreich's term for the variety of services and products aimed at self-motivation, such as motivational speakers, seminars, guides, books, calendars, and workplace posters.

100. Cerulo, *Never Saw It Coming*, 64.

101. Respondents 9565610, 9580101, 9692814, 9566158.

102. For an insightful comparison between pre-9/11 optimism and postrecession perspectives, see Pew Research Center for the People and the Press, "Public Sees a Future Full of Promise and Peril," June 22, 2010, http://www.people-press.org/2010/06/22/public-sees-a-future-full-of-promise-and-peril.

103. See Kari Marie Norgaard, *Living in Denial: Climate Change, Emotions, and Everyday Life* (Cambridge, MA: MIT Press, 2011), 97–135

104. Respondent 9597368. See also Ehrenreich, *Bright-Sided*, 8.

105. Cerulo, *Never Saw It Coming*, 65.

106. Respondents 9601195, 9601157, 9580171, 9585039, 9570884.

107. Richard McNeill Douglas, "The Ultimate Paradigm Shift: Environmentalism as Antithesis to the Modern Paradigm of Progress," in *Future Ethics: Climate Change and Apocalyptic Imagination*, ed. Stefan Skrimshire (New York: Continuum: 2010), 197–218, 206. See also Riley E. Dunlap and K. D. Van Liere, "The 'New Environmental Paradigm': A Proposed Measuring Instrument and Preliminary Results," *Journal of Environmental Education* 9, no. 4 (1978): 10–19.

108. The phrase *dominant social paradigm*, or what some might call *ideology*, was coined in Dunlap and VanLiere, "The 'New Environmental Paradigm.'"

109. Respondents 9590707, 9563017, 9577121, 9577121, 9563747.

110. Evgeny Morozov, *To Save Everything, Click Here: The Folly of Technological Solutionism* (New York: PublicAffairs, 2013), 5. This question is also addressed, in a slightly different way, by sociologists of risk and disasters, with scholars like Ulrich Beck and Kai Erikson remaining wary of relying on technology and experts and scholars such as Anthony Giddens, Penelope Canan, and Nancy Reichman retaining much more confidence.

111. Respondent 9565610.

112. Respondent 9600939. As many commentators have noted, the last national politician or major public figure to criticize Americans' lifestyles

and call for sacrifice was Jimmy Carter, in his infamous "crisis of confidence" speech of 1979. Although the initial reaction to the speech was actually positive, the success of Carter's political opponents (both Republicans and Democrats) in tying the assertion of limits to Carter's political incompetence and impotence provided a lasting lesson about the demand for optimism in public rhetoric. See Kevin Mattson, *What the Heck Are You Up to, Mr. President? Jimmy Carter, America's "Malaise," and the Speech That Should Have Changed the Country* (New York: Bloomsbury, 2009); and my discussion of the legacy of 1970s limits discourse in chapter 2.

113. Respondents 9559395, 9589722, 9574202.

CHAPTER TWO

1. Jeffrey Ball, "As Prices Soar, Doomsayers Provoke Debate on Oil's Future," *Wall Street Journal*, September 21, 2004, A1; Colin J. Campbell, "The End of Cheap Oil," *Scientific American* 278, no. 3 (1998): 60–65, and *The Coming Oil Crisis*.

2. Leonardo Maugeri, "Squeezing More Oil from the Ground," *Scientific American* 301, no. 4 (2009): 56–63. For one of Campbell's interviews, see Michael C. Ruppert, "Colin Campbell on Oil: Perhaps the World's Foremost Expert on Oil and the Oil Business Confirms the Ever More Apparent Reality of the Post–9-11 World," From the Wilderness, October 23, 2002, http://www.fromthewilderness.com/free/ww3/102302_campbell .html. Colin Campbell, personal communication, February 6, 2010.

3. Doris Leblond, "ASPO Sees Conventional Oil Production Peaking by 2010," *Oil and Gas Journal*, June 30, 2003, 28.

4. The Coming Global Oil Crisis, http://www.hubbertpeak.com; Oil Analytics, http://www.oilanalytics.net; Die Off, http://www.dieoff.com. In this section, I trace the origins and development of Web sites using the Internet Archive (also known as the "Wayback Machine"), http://www.archive .org/index.php.

5. On social movements and television, see Carol A. Stabile, *White Victims, Black Villains: Gender, Race and Crime News in US Culture* (New York: Routledge, 2006).

6. Henry Jenkins, "The Cultural Logic of Media Convergence," *International Journal of Cultural Studies* 7, no. 33 (2004): 35. For a fuller description of convergence culture, see Henry Jenkins, *Convergence Culture: Where Old and New Media Collide* (New York: New York University Press, 2006).

7. Bruce Stanley, "Oil Experts Draw Fire for Warning," *Dubuque (IA) Telegraph Herald*, May 27, 2002, B5. Simmons, a speaker at almost every ASPO conference until his passing in 2010, was noteworthy not so much for his actual expertise as for the imprimatur of legitimacy he provided. In this 2002 article he is described as one of many advisers to the candidate

George W. Bush on energy issues, but by the end of the decade many peakists described him as something akin to President Bush's right-hand man.

8. For example, Michael C. Lynch, "Petroleum Resources Pessimism Debunked in Hubbert Model and Hubbert Modelers' Assessment," *Oil and Gas Journal*, July 14, 2003, 38, http://www.ogj.com/articles/print/volume-101/issue-27/special-report/petroleum-resources-pessimism-debunked-in-hubbert-model-and-hubbert-modelers-assessment.html; and John H. Wood, Gary R. Long, and David F. Morehouse, "World Conventional Oil Supply Expected to Peak in 21st Century," *Offshore*, April 1, 2003, http://www.offshore-mag.com/articles/print/volume-63/issue-4/technology/world-conventional-oil-supply-expected-to-peak-in-21st-century.html.

9. Peak Oil News and Message Boards, http://www.peakoil.com; The Energy Bulletin, https://www.energybulletin.net (now defunct); The Oil Drum, http://www.theoildrum.com; Life After the Oil Crash, http://www.lifeaftertheoilcrash.com (now defunct).

10. This process is well documented by sociologists and journalists. See, e.g., Peter L. M. Vasterman, "Media-Hype: Self-Reinforcing News Waves, Journalistic Standards and the Construction of Social Problems," *European Journal of Communication* 20, no. 4 (2005): 508–30.

11. Daniel Yergin, "Imagining a $7-a-Gallon Future," *New York Times*, April 4, 2004, WK1; Terence Corcoran, "The 'Peak Oil' Cult," *National Post* (Don Mills, ON), October 12, 2004, FP15; Michael R. Duffey, "Prognosticating Oil Supplies: Lower Reserves and Higher Prices Ahead?" *Washington Times*, November 3, 2004, A23; Paul Roberts, "Cheap Oil, the Only Oil That Matters, Is Just about Gone," *Harper's*, August 2004, 71–72.

12. In 2003, the major peak oil book available was Richard Heinberg's *The Party's Over*. Thom Hartman, *The Last Hours of Ancient Sunlight: The Fate of the World and What We Can Do Before It's Too Late* (New York: Three Rivers, 2004); Darley, *High Noon for Natural Gas*; Goodstein, *Out of Gas*; Sonia Shah, *Crude: The Story of Oil* (Toronto: Seven Stories, 2004); Campbell, *The Coming Oil Crisis*; Ronald R. Cooke, *Oil, Jihad and Destiny* (New York: Opportunity Analysis, 2004); Michael Ruppert, *Crossing the Rubicon: The Decline of the American Empire at the End of the Age of Oil* (New York: New Society Publishers, 2004); Deffeyes, *Beyond Oil*; Michael T. Klare, *Blood and Oil: The Dangers and Consequences of America's Growing Dependence on Petroleum* (New York: Henry Holt, 2004); and Richard Heinberg, *Powerdown: Options and Actions for a Post-Carbon World* (West Hoathly: Clairview, 2004).

13. Peak Oil News and Message Boards poll entitled "I First Learned about Peak Oil," http://peakoil.com/modules.php?name=Surveys&op=results&pollID=2.

14. Pew Internet and American Life Project Surveys, "Digital Differences,"

April 13, 2012, http://pewinternet.org/Reports/2012/Digital-differences/ Main-Report/Internet-adoption-over-time.aspx; Thom File and Camille Ryan, "Computer and Internet Use in the United States: 2013," U.S. Census Bureau, November 2014, http://www.census.gov/content/dam/Census/ library/publications/2014/acs/acs-28.pdf.

15. By 2008, the stream of peak oil evangelical works had turned into a flood. Some of the most significant works are Tom Mast, *Over a Barrel: A Simple Guide to the Oil Shortage* (New York: Hayden, 2005); Matthew R. Simmons, *Twilight in the Desert: The Coming Saudi Oil Shock and the World Economy* (Hoboken, NJ: Wiley, 2005); Peter Tertzakian, *A Thousand Barrels a Second: The Coming Oil Break Point and the Challenges Facing an Energy Dependent World* (New York: McGraw-Hill, 2006); Edwin Black, *Internal Combustion: How Corporations and Governments Addicted the World to Oil and Derailed the Alternatives* (New York: St. Martin's, 2006); David Strahan, *The Last Oil Shock: A Survival Guide to the Imminent Extinction of Petroleum Man* (New York: John Murray, 2007); and Michael T. Klare, *Rising Powers, Shrinking Planet: The New Geopolitics of Energy* (New York: Henry Holt, 2008).

16. *The End of Suburbia: Oil Depletion and the Collapse of the American Dream* (Electric Wallpaper Co., 2004); Natalie Cavanor, "Running on Empty; Two Video Producers Paint a Bleak Picture of Oil Wars and the End of the American Suburbs," *New York Times*, March 13, 2005, LI1.

17. Caryl Johnston, *After the Crash: An Essay-Novel of the Post Hydrocarbon Age* (self-published, 2004); Urstadt, "Imagine There's No Oil."

18. James Howard Kunstler, "Wake Up, America; We're Driving toward Disaster," *Washington Post*, May 28, 2005, B3; Joe Raedle, "Are We There Yet? Oil Joyride May Be Over," *USA Today*, May 28, 2005, B03; Matthew DeBord, Review of *$20 Per Gallon* by Christopher Steiner, *Los Angeles Times*, May 26, 2009, http://articles.latimes.com/2009/jul/26/entertainment/ca-christopher -steiner26; Justin Fox, "Peak Possibilities," *Time*, November 21, 2007, http:// content.time.com/time/magazine/article/0,9171,1686824,00.html.

19. Alex Williams, "Duck and Cover: It's the New Survivalism," *New York Times*, April 6, 2008, ST-1; Peter Maass, "The Breaking Point," *New York Times*, August 21, 2005, E30; Joseph Nocera, "On Oil Supply, Opinions Aren't Scarce," *New York Times*, September 10, 2005, C1; Paul Brown, "The Deathwatch for Cheap Oil," *New York Times*, October 14, 2006, C5; Roger Lowenstein, "What Price Oil?" *New York Times*, October 19, 2008, 46; Jon Mooallem, "The End Is Near! (YAY!)," *New York Times*, April 19, 2008, MM28; Lynch, "'Peak Oil' Is a Waste of Energy."

20. Philip Niemeyer, "Picturing the Past Ten Years," *New York Times*, December 27, 2009, http://www.nytimes.com/interactive/2009/12/27/opinion/ 28opchart.html.

21. Samantha Gross, "Energy Fears Looming, New Survivalists Prepare: Too Late to Save the Planet, They Say, So They Focus on Saving Themselves,"

MSNBC.com, May 24, 2006, http://www.msnbc.msn.com/id/24808083. The Drudge Report linked directly to the Associated Press News Web site in its "May 25, 2008 00:45:23 GMT edition," http://www.drudgereport.com.

22. The connection between Gross's article and the site's membership was confirmed by an administrator of Peak Oil News and Message Boards. "Aaron," personal communication, January 28, 2009.

23. "Oil Prices Soar, Stocks Plummet: How Bad Is Our Economy? Hillary Clinton Accepts Defeat," CNN *Saturday Morning News*, June 7, 2008, transcript at http://transcripts.cnn.com/TRANSCRIPTS/0806/07/smn.02.html.

24. As these examples show, the traditional media retain their power to spread information in our convergence culture, but they do not define the terms of debate.

25. See John W. Western, *Selling Intervention and War: The Presidency, the Media, and the American Public* (Baltimore: Johns Hopkins University Press, 2005); and W. Lance Bennett, Regina G. Lawrence, and Steven Livingston, *When the Press Fails: Political Power and the News Media from Iraq to Katrina* (Chicago: University of Chicago Press, 2007).

26. Gary Leupp, "The Weekend the World Said No to War: Notes on the Numbers," Counterpunch, February 25, 2003, http://www.counterpunch.org/2003/02/25/the-weekend-the-world-said-no-to-war-notes-on-the-numbers.

27. For example, one CBS report juxtaposed the following aspects of an antiwar event: "'Can you justify blood for oil?' read a sign held by 14-year old Marianna Daniels at a rally in Madison, Wis. The New York rally was opened by singer Richie Havens performing 'Freedom,' just as he did 34 years earlier at the original Woodstock Festival. Speakers included Susan Sarandon, Harry Belafonte and Pete Seeger." By linking a potentially insightful critique of American militarism to its author's age, musical entertainment, and the Woodstock Festival, the article implicitly minimized the claim of petroimperialism. Sue Chan, "Massive Anti-War Outpouring," CBS News, February 16, 2003, http://www.cbsnews.com/news/massive-anti-war-outpouring.

28. Kevin Phillips, *American Theocracy: The Peril and Politics of Radical Religion, Oil, and Borrowed Money in the 21st Century* (New York: Viking, 2006), 69.

29. Bob Woodward, "Greenspan Is Critical of Bush in Memoir: Former Fed Chairman Has Praise for Clinton," *Washington Post*, September 16, 2007, A1.

30. Respondents 9573008, 9589283, 9683776.

31. Although gasoline prices generally reflect the cost of a barrel of oil, the US Department of Energy estimates that the price of crude has only a 67 percent stake in determining the price of gasoline at the pump. The rest reflects the costs of refining, distribution and marketing by corporations, state and federal taxes, and station markup. Since station managers themselves must predict what the price will be before they purchase drums of gasoline, the price of gasoline is, at times, temporarily out of step with the price of oil.

32. See, e.g., Heinberg, *The Party's Over*; Kunstler, *The Long Emergency*; and Roberts, *The End of Oil*.
33. David S. Painter, "Oil and the American Century," *Journal of American History* 99, no. 1 (2012): 24–39, 26.
34. Yergin, *The Prize*, 208.
35. Painter, "Oil and the American Century," 27; Yergin, *The Prize*, 334–67.
36. Comment by Britain's Lord Curzon on the Allies' victory, Inter-Allied Petroleum Conference at Lancaster House, 1918, quoted in Yergin, *The Prize*, 183.
37. Yergin, *The Prize*, 15; Henry Luce, "The American Century," *Life*, February 17, 1941, 61.
38. Natasha Zaretsky, *No Direction Home: The American Family and the Fear of National Decline, 1968–1980* (Chapel Hill: University of North Carolina Press, 2007), 89.
39. Lizabeth Cohen, *A Consumer's Republic: The Politics of Mass Consumption in Postwar America* (New York: Vintage, 2003). Typical of most social and political histories in its marginalization of energy as a factor in American history, Cohen's influential work barely mentions petroleum.
40. Brian C. Black, "Oil for Living: Petroleum and American Conspicuous Consumption," *Journal of American History* 99, no. 1 (2012): 40–50, 50.
41. Painter, "Oil and the American Century," 25.
42. Priest, "The Dilemmas of Oil Empire," 236.
43. Ibid., 245. See also Marshall I. Goldman, *Petrostate: Putin, Power, and the New Russia* (Oxford: Oxford University Press, 2008).
44. The average price of gasoline rose from $1.27 in January 2000 to a high of $4.11 in July 2008, while the average barrel price rose from $22.68 to a high of $133.60.
45. At the time, Venezuela supplied the United States with over 800,000 barrels per day, and its president, Hugo Chavez, was an outspoken critic of President Bush. US Energy Information Administration, "U.S. Imports by Country of Origin," December 30, 2014, http://www.eia.gov/dnav/pet/pet_move_impcus_a2_nus_epc0_im0_mbblpd_a.htm.
46. Insofar as Hurricane Katrina was a major factor in the increase of gas prices, one could potentially subsume gas prices under the category of the environment in this section. However, to do so would present at least two problems. First, it would suggest that gas prices are as natural as the weather, the result of either an environmental determinism or a naturalization of the free market principles (supply and demand) that are assumed to control commodity prices. While environmental factors clearly have a major impact on the cost of energy production, there are many other factors, which this section tries to consider. The tendency to subsume energy (particularly oil) under the general category of the environment can be expedient, but it can also be misleading.
47. Respondent 9592346.

48. Sean Alfano, "Poll: Gas Prices Affecting Habits," CBS News, September 1, 2005, http://www.cbsnews.com/news/poll-gas-prices-affecting-habits.

49. Respondents 9569670, 9564499, 9569883.

50. Twenty-seven percent answered "a fair amount," 21 percent "only a little," and 12 percent "not at all." "In U.S., Concerns about Global Warming Stable at Lower Levels," Gallup, March 14, 2011, http://www.gallup.com/poll/146606/concerns-global-warming-stable-lower-levels.aspx.

51. A number of surveys found that the film significantly increased its audience's concern about climate change. See, e.g., Andrew Balmford, Linda Birkin, Andrea Manica, Lesley Airey, Amy Oliver, and Judith Schleicher, "Hollywood, Climate Change, and the Public," *Science* 305, no. 5691 (2004): 1713.

52. Zogby International, "Americans Link Katrina, Global Warming," 2006, http://www.zogby.com/News/ReadNews.dbm?ID=1161 (now defunct).

53. Michael Ziser and Julie Sze, "Climate Change, Environmental Aesthetics, and Global Environmental Justice Cultural Studies," *Discourse* 29, nos. 2/3 (2007): 384–410, 402.

54. *An Inconvenient Truth*, Davis Guggenheim (2006).

55. Nielsen, "Global Consumers Vote Al Gore, Oprah Winfrey and Kofi Annan Most Influential to Champion Global Warming Cause: Nielsen Survey," July 2, 2007, http://www.eci.ox.ac.uk/news/press-releases/070703pr-climatechamps-world.pdf. However, Gore's association with the issue allowed conservative climate change deniers to politicize it by claiming that climate science was simply liberal propaganda.

56. Gary Langer, "Poll: Public Concern on Warming Gains Intensity," ABC News, March 26, 2006, http://abcnews.go.com/Technology/GlobalWarming/story?id=1750492&page=1.

57. Interview with Steve Allen, November 11, 2009; interview with Bill Fickas, November 12, 2009; respondent 9569902.

58. Klaus Mann, preface to *America*, by Franz Kafka (New York: Schocken, 1946), xxvi.

59. John F. Kennedy, speech given in Washington, DC, on January 20, 1961.

60. On the idea of progress throughout history, see J. B. Bury, *The Idea of Progress: An Inquiry into Its Origins and Growth* (Cambridge: Cambridge University Press, 1920).

61. J. R. McNeill, *Something New under the Sun: An Environmental History of the Twentieth-Century World* (New York: Norton, 2000), 336.

62. Robert M. Collins, *More: The Politics of Economic Growth in the Postwar World* (New York: Oxford University Press, 2000), 2–3; Frederick Jackson Turner, *The Significance of the Frontier in American History*, Great Ideas (Harmondsworth: Penguin, 2008).

63. Russell Frank Wiegley, *The American Way of War: A History of United States Military and Policy* (New York: Macmillan, 1973), 146. Even more broadly, speaking of a national economy before the 1940s is somewhat anachro-

nistic. As Timothy Mitchell argues, no political economist referred to an object called *the economy* prior to the 1930s. The emergence of the notion of the economy was dependent on two factors. First was the expanded national administrative machinery that developed only with the New Deal. Second was a consensus in the field of economics about how to measure a national economy—or whether that was even possible or desirable. In the 1920s, economists such as Richard T. Ely and Thorstein Veblen "wanted economics to start from natural resources and flows of energy," while a competing group wanted "to organize the discipline around the study of prices and the flows of money." Mitchell, *Carbon Democracy*, 132. The second group won out, and as a result the national economy has been measured by the amount of times that money changes hands and questions of energy and natural resources have, for the most part, been ignored.

64. Godfrey Hodgson, *America in Our Time: From World War II to Nixon—What Happened and Why* (Garden City, NJ: Doubleday, 1976), 48.

65. Collins, *More*, 21.

66. Harry S. Truman, "Annual Message to the Congress on the State of the Union" (January 5, 1949), in *Public Papers of the Presidents of the United States: Harry S. Truman* (Washington, DC: US Government Printing Office, 1964), 1–7.

67. See Richard Slotkin, *Gunfighter Nation: The Myth of the Frontier in Twentieth-Century America* (New York: HarperPerennial, 1993), 347–533.

68. Hansen quoted in Alan Brinkley, "The New Deal and the Idea of the State," in *The Rise and Fall of the New Deal Order, 1930–1980*, ed. Steve Fraser and Gary Gerstle (Princeton, NJ: Princeton University Press, 1989), 85–121, 108.

69. Cohen, *A Consumer's Republic*, 119.

70. David Potter, *People of Plenty: Economic Abundance and the American Character* (Chicago: University of Chicago Press, 1954), 70.

71. William H. Whyte, *The Organization Man* (New York: Simon & Schuster, 1956), 14.

72. On Eisenhower and growth, see Collins, *More*, 43–45.

73. Ibid., 52.

74. Hodgson, *America in Our Time*, 52.

75. While it is tempting to think of growth as a specifically capitalist goal, Communist regimes, particularly that in the Soviet Union, practiced their own form of competitive growthmanship. Indeed, as McNeill has argued, while "Communism aspired to become the universal creed of the twentieth century," in reality "a more flexible and seductive religion succeeded where communism failed: the quest for economic growth." McNeill, *Something New under the Sun*, 336.

76. Lawrence R. Samuel, *Future: A Recent History* (Austin: University of Texas Press, 2009), 108.

77. James T. Patterson, *Grand Expectations: The United States, 1945–1974* (New York: Oxford University Press, 1996), vii.

78. While these changes and the related sense of growing possibilities affected most Americans, some had more possibilities than others. While people of color, gays and lesbians, and poor Americans still suffered from wide-spread discrimination, oppression, and disenfranchisement, the sense of growth and possibility may have played a role in the political movements that emerged in the 1960s.

79. See Samuel, *Future*, 77–108. Radebaugh was an illustrator whose sleek, airbrushed advertisements for companies such as Coca-Cola and United Airlines had a powerful influence on contemporary future imaginaries. His syndicated cartoon strip *Closer Than We Think* (1958–62) presented the mass attainment of a better life through technological progress, through cartoons demonstrating breakthroughs such as "Push-Button Education," "Robot Warehouses," and of course the flying automobile. See Matt Novak, "Before the Jetsons, Arthur Radebaugh Illustrated the Future," *Smithsonian*, April 2012, 30. Alvin Toffler was a well-known futurist whose articles, such as "The Future as a Way of Life" (*Horizon*, Summer 1965, 108–16), and books, such as *Future Shock* (New York: Random House, 1990), had a wide cultural resonance.

80. Black, "Oil for Living," 43.

81. Dwight D. Eisenhower, Annual Message to the Congress on the State of the Union, January 6, 1955, http://www.eisenhower.archives.gov/all_about_ike/speeches/1955_state_of_the_union.pdf.

82. Appearing frequently at the time, the term *golden age* was used by no less a historian than Eric Hobsbawm to describe life in the "developed capitalist countries" from roughly 1945 to 1973, and Hobsbawm notes that other historians have chosen similar terms—the French call this period their "thirty glorious years" (*les trente glorieuses*) and the British "the Golden Age of the Anglo-Americans." Eric Hobsbawm, *The Age of Extremes: A History of the World, 1914–1991* (New York: Vintage, 1994), 258.

83. Collins, *More*, 2.

84. Imre Szeman, "The Cultural Politics of Oil: On Lessons of Darkness and Black Sea Files," *Polygraph* 22 (2010): 34; Michael G. Ziser, "Home Again: Peak Oil, Climate Change, and the Aesthetics of Transition," in *Environmental Criticism for the Twenty-First Century*, ed. Stephanie LeMenager, Teresa Shewry, and Ken Hiltner (New York: Routledge, 2011), 181–95, 183.

85. See Brian M. Fagan, *The Great Journey: The Peopling of Ancient America* (Gainesville: University Press of Florida, 2003), 190–95; Diamond, *Collapse*, 9–10; and Jared Diamond, *Guns, Germs and Steel: The Fates of Human Societies* (New York: Norton, 1999), 42–47.

86. Clive Ponting, *A Green History of the World: The Environment and the Collapse of Great Civilizations* (New York: Penguin, 1991), 289.

87. McNeill, *Something New under the Sun*, 15.

88. Ibid.
89. Thomas Robertson, *The Malthusian Moment: Global Population Growth and the Birth of American Environmentalism* (New Brunswick, NJ: Rutgers University Press, 2012).
90. Paul Ehrlich, *The Population Bomb* (New York: Ballantine, 1968); Garrett Hardin, "Tragedy of the Commons," *Science* 162 (3859): 1243–48.
91. See Elaine Tyler May, *America and the Pill: A History of Promise, Peril, and Liberation* (New York: Basic, 2010), 35–56.
92. John S. Dryzek, *The Politics of the Earth: Environmental Discourses* (Oxford: Oxford University Press, 2012), 26.
93. *Congressional Record*, September 12, 1969. The article was "Eco-Catastrophe," originally published in *Ramparts*, September 1969, 24–28.
94. Amitai Etzioni, "America's Project" (1979), quoted in Collins, *More*, 165.
95. Max Lerner, "Just Imagine! We All Can Avoid a Certain Doomsday," *Los Angeles Times*, March 10, 1972, C9.
96. Donella H. Meadows, Dennis L. Meadows, Jørgen Randers, and William W. Behrens III, *The Limits to Growth* (New York: Universe, 1972).
97. *Business Week*, March 11, 1972, 98.
98. Ronald Kotulak, "Fuel, Resources Dwindling: Can America Survive the 20th Century?" *Chicago Tribune*, March 4, 1973, 1.
99. Gladwin Hill, "Nixon Aide Asks Wide Debate on Desirability of U.S. Growth," *New York Times*, March 30, 1972, 19.
100. Respondent 9590491.
101. As recent events in the United States have shown, the real motivation for war often remains a mystery, but many historians claim that for some OPEC nations a secondary explanation was the desire to stabilize their real incomes by raising oil prices, which would become far easier with increased leverage in the wake of asserting their power via an embargo. See Yergin, *The Prize*, 589–632.
102. Peter N. Carroll, *It Seemed Like Nothing Happened: America in the 1970s* (New Brunswick, NJ: Rutgers University Press, 2000), 118.
103. E. F. Schumacher, *Small Is Beautiful: A Study of Economics as If People Mattered* (London: Blond & Briggs, 1973).
104. Riley E. Dunlap and William R. Carton Jr., "Struggling with Human Exemptionalism: The Rise, Decline and Revitalization of Environmental Sociology," *American Sociologist* 25, no. 1 (1994): 8.
105. Jefferson Cowie, *Stayin' Alive: The 1970s and the Last Days of the Working Class* (New York: New Press, 2010), 317.
106. Frederick Buell, *From Apocalypse to Way of Life: Environmental Crisis in the American Century* (New York: Routledge, 2004), 20.
107. Josh Eastin, Reiner Grundmann, and Andeem Prakash, "The Two Limits Debates: 'Limits to Growth' and Climate Change," *Futures* 43, no. 1 (2011): 16–26, 18. For example, *Limits to Growth* estimated (in 1972) that world lead reserves would be entirely depleted in twenty-six years, mercury in

thirteen years, silver in sixteen years, tin in seventeen years, natural gas in thirty-eight years, and petroleum in thirty-one years.

108. Robertson, *The Malthusian Moment*, 11.
109. Ronald Reagan, Inaugural Address, January 20, 1981, http://www .presidency.ucsb.edu/ws/?pid=43130.
110. See Jeffrey K. Stine, "Natural Resources and Environmental Policy," in *The Reagan Presidency: Pragmatic Conservatism and Its Legacies*, ed. W. Elliott Brownlee and Hugh Davis Graham (Lawrence: University Press of Kansas, 2003), 233–58.
111. See Frederick H. Buttel, Ann P. Hawkins, and Alison G. Power, "From Limits to Growth to Global Change: Constraints and Contradictions in the Evolution of Environmental Science and Ideology," *Global Environmental Change* 1, no. 1 (1990): 57–66.
112. Mulligan, "Energy, Environment, and Security," 84.
113. US Department of Labor, Bureau of Labor Statistics, Consumer Price Index-Average Price Data, http://data.bls.gov/cgi-bin/surveymost?ap. Adjusted for inflation.
114. Yergin, *The Prize*, 715–44.
115. Collins, *More*, 227.
116. Respondent 9688522.
117. Exxon Mobil Corporation, "Limits to Growth?" *New York Times*, July 25, 2002, A17.
118. Philip Stott, "Hot Air + Flawed Science = Dangerous Emissions," *Wall Street Journal*, April 2, 2001, A22.
119. Bill McKibben, *Eaarth: Making a Life on a Tough New Planet* (New York: Henry Holt, 2010), 27–33.
120. Sandra Steingraber, *Living Downstream: An Ecologist's Personal Investigation of Cancer and the Environment* (1997), 2nd ed. (Philadelphia: Da Capo, 2010), 117.
121. Kharecha A. Pushker and James E. Hansen, "Implications of 'Peak Oil' for Atmospheric CO2 and Climate" (New York: NASA Goddard Institute for Space Studies and Columbia University Earth Institute, n.d.), http://arxiv .org/ftp/arxiv/papers/0704/0704.2782.pdf.
122. Christian Parenti, *Tropic of Chaos: Climate Change and the New Geography of Violence* (New York: Nation, 2011), 7.
123. See, e.g., A. P. Ballantyne, C. B. Alden, J. B. Miller, P. P. Tans, and J. W. C. White, "Increase in Observed Net Carbon Dioxide Uptake by Land and Oceans during the Past 50 Years," *Nature* 488, no. 7409 (2012): 70–72; Megan I. Saunders, Javier Leon, Stuart R. Phinn, David P. Callaghan, Katherine R. O'Brien, Chris M. Roelfsema, Catherine E. Lovelock, Mitchell B. Lyons, and Peter J. Mumby, "Coastal Retreat and Improved Water Quality Mitigate Losses of Seagrass from Sea Level Rise," *Global Change Biology* 19, no. 8 (2013): 2569–83.
124. Bill McKibben, "Global Warming's Terrifying New Math," *Rolling Stone*,

July 19, 2012, http://www.rollingstone.com/politics/news/global-warmings
-terrifying-new-math-20120719.
125. Douglas, "The Ultimate Paradigm Shift," 200.

CHAPTER THREE

1. Wellman, "Little Boxes, Glocalization, and Networked Individualism."
2. Sherry Turkle, *Alone Together: Why We Expect More from Technology and Less from Each Other* (New York: Basic, 2011).
3. Mooallem, "The End Is Near! (Yay!)," MM28.
4. "Transition US," http://www.transitionus.org/transition-towns.
5. "Towns Banking Their Own Currency," BBC News, April 2, 2008, http://news.bbc.co.uk/2/hi/uk_news/wales/south_west/7326212.stm.
6. These articles have received special attention in British scholarship. See, e.g., North, "Eco-Localisation as a Progressive Response"; Bailey, Hopkins, and Wilson, "Some Things Old, Some Things New"; J. P. Evans, "Resilience, Ecology and Adaptation in the Experimental City," *Transactions of the Institute of British Geographers* 36, no. 2 (2011): 223–37; and Smith, "The Transition Town Network."
7. Interview with Gaea Swinford, June 6, 2011.
8. Mooallem, "The End Is Near! (Yay!)," MM28.
9. Respondents 9590659, 9591372, 9584855, 9602174.
10. This analysis is intended not to detract from the Transition Initiative movement, which is certainly vibrant in some areas, but to point out that in the United States it can function as a Potemkin village for communitarian-minded observers.
11. Heinberg, *The Party's Over*, 214; respondents 9567268, 9567956, 9592645.
12. Respondents 9563196, 9559395, 9562254, 9568949, 9565756, 9682014, 9591372, 9592893.
13. Adam Rome, "'Give Earth a Chance': The Environmental Movement and the Sixties," *Journal of American History* 90, no. 2 (2003): 544.
14. Paul Lichterman, *The Search for Political Community: American Activists Reinventing Commitment* (Cambridge: Cambridge University Press, 1996), 6.
15. Felicia Wu Song, *Virtual Communities: Bowling Alone, Online Together* (New York: Peter Lang, 2009), 115.
16. Lichterman, *The Search for Political Community*, 6, 19. See also, e.g., Sulamith H. Potter, "The Cultural Construction of Emotion in Rural Chinese Social Life," *Ethos* 16, no. 2 (1988): 181–208.
17. Amitai Etzioni, ed., *The Essential Communitarian Reader* (Lanham, MD: Rowman & Littlefield, 1998).
18. David Kirby and David Boaz, *The Libertarian Vote in the Age of Obama*, Policy Analysis no. 658 (Washington, DC: Cato Institute, January 21, 2010). Although, as a libertarian think tank, the Cato Institute, which publishes the series Policy Analysis, does have a stake in presenting libertarianism

as a political philosophy with wide appeal, the methodology used in this article appears to be without bias.

19. Nate Silver, "Poll Finds a Shift toward More Libertarian Views," *New York Times*, June 20, 2011, http://fivethirtyeight.blogs.nytimes.com/2011/06/20/poll-finds-a-shift-toward-more-libertarian-views.

20. William E. Hudson, *The Libertarian Illusion: Ideology, Public Policy, and the Assault on the Common Good* (Washington, DC: CQ Press, 2008), 1–21. While Hudson is critical of libertarianism, his opening description of its basic tenets is relatively even-handed. A more partisan and positive description can be found in Brian Doherty, *Radicals for Capitalism: A Freewheeling History of the Modern American Libertarian Movement* (New York: PublicAffairs, 2007).

21. See Lisa McGirr, *Suburban Warriors: The Origins of the New American Right* (Princeton, NJ: Princeton University Press, 2001); Kevin M. Kruse, *White Flight: Atlanta and the Making of Modern Conservatism* (Princeton, NJ: Princeton University Press, 2005); Thomas J. Sugrue, *The Origins of the Urban Crisis: Race and Inequality in Postwar Detroit* (Princeton, NJ: Princeton University Press, 1996); and Joseph Crespino, *In Search of Another Country: Mississippi and the Conservative Counterrevolution* (Princeton, NJ: Princeton University Press, 2007). Indeed, some historians argue that, instead of seeing conservatism as a social movement that wrested power from normative New Deal liberalism, we should view the ascendence of that liberalism itself as a mere interregnum in the larger history of what has always been an essentially conservative country.

22. See Ronald P. Formisano, *Boston against Busing: Race, Class, and Ethnicity in the 1960s and 1970s* (Chapel Hill: University of North Carolina Press, 1991); David Farber and Jeff Roche, *The Conservative Sixties* (New York: Peter Lang, 2003); Rebecca E. Klatch, *A Generation Divided: The New Left, the New Right, and the 1960s* (Berkeley and Los Angeles: University of California Press, 1999); Ronnee Schreiber, *Righting Feminism: Conservative Women and American Politics* (New York: Oxford University Press, 2008); Donald T. Critchlow, *Phyllis Schlafly and Grassroots Conservatism: A Woman's Crusade* (Princeton, NJ: Princeton University Press, 2005); and Robert O. Self, *American Babylon: Race and the Struggle for Postwar Oakland* (Princeton, NJ: Princeton University Press, 2003).

23. See Steven P. Miller, *Billy Graham and the Rise of the Republican South* (Philadelphia: University of Pennsylvania Press, 2009); and Julian E. Zelizer and Bruce J. Schulman, eds., *Rightward Bound: Making America Conservative in the 1970s* (Cambridge, MA: Harvard University Press, 2008).

24. Some scholars, such as Hyrum Lewis, have argued that the definition of what count as conservative ideas has changed so much from the 1950s to the 2010s that conservatism cannot be defined as one coherent movement. While I second this caution against essentializing, I believe that the

Right has had a number of core principles that have remained relatively static, even if they have been more or less prominent over time. Hyrum Lewis, "Historians and the Myth of American Conservatism," *Journal of the Historical Society* 12, no. 1 (2012): 27–45.

25. Although opposition to big government has often been claimed as a central tenet of conservatism, numerous critics have pointed out that, when in control, few conservatives have (until recently, perhaps) truly attempted to dismantle the welfare state and, therefore, that the "liberal-conservative divide is not so much over whether or not to limit government, but in what ways to limit government—liberals favor expansion of the welfare state, and conservatives favor expansion of the military-morality state" (ibid., 34).

26. See, e.g., Formisano, *Boston against Busing*; Catherine E. Rymph, *Republican Women: Feminism and Conservatism from Suffrage to the New Right* (Chapel Hill: University of North Carolina Press, 2006); Crespino, *In Search of Another Country*; Kruse, *White Flight*; and Joseph E. Lowndes, *From the New Deal to the New Right: Race and the Southern Origins of Modern Conservatism* (New Haven, CT: Yale University Press, 2008). For trenchant historical critiques of the myth of traditional American family values, see Stephanie Coontz, *The Way We Never Were: American Families and the Nostalgia Trap* (New York: Basic, 2000); and Matthew D. Lassiter, "Inventing Family Values," in Schulman and Zelizer, eds., *Rightward Bound*, 13–28.

27. See Julian Zelizer, "Reflections: Rethinking the History of American Conservatism," *Reviews in American History* 38, no. 2 (2010): 367–92, 380. The biggest victory for liberals over the last two decades has been, without question, the expansion of various human rights to Americans of all genders and sexualities.

28. While this broad definition of *neoliberalism* would be accepted by most scholars, there is a great deal of disagreement about exactly how to define it. For a diversity of perspectives, see Philip Mirowski and Dieter Plehwe, eds., *The Road from Mont Pèlerin: The Making of the Neoliberal Thought Collective* (Cambridge, MA: Harvard University Press, 2009).

29. Libertarian ideas go under different names outside the United States—they are generally called *liberal*—but the difference here is not solely in terms of language. I explore the exceptional popularity of libertarian ideas in the United States over the next few pages.

30. Seymour Lipset, *American Exceptionalism: A Double-Edged Sword* (New York: Norton, 1996), 46.

31. *New York Times*, February 3, 1971, 47.

32. Hudson, *The Libertarian Illusion*, 24–25.

33. Daniel Rodgers, *The Age of Fracture* (Cambridge, MA: Harvard University Press, 2011), 8, 198.

34. Teles and Kenney, "Spreading the Word," 168.

35. Bill Clinton, "Address Before a Joint Session of the Congress on the State of the Union," January 23, 1996, http://www.presidency.ucsb.edu/ws/?pid =53091.

36. Steven R. Weisman, "Reagan Tiptoes around Some Economic Liabilities," *New York Times*, September 26, 1982, sec. 4, p. 4.

37. These calls for austerity have come primarily from Tea Party conservatives. While they may not be popular in that they do not have the broad support of most Americans, they are, indeed, populist, coming from not only political pundits or the moneyed elite but also many average Americans. See Vanessa Williamson, Theda Skocpol, and John Coggin, *The Tea Party and the Remaking of Republican Conservatism* (New York: Oxford University Press, 2012), esp. chap. 2; and Binyamin Appelbaum and Robert Gebeloff, "Even Critics of Safety Net Increasingly Depend on It," *New York Times*, February 12, 2012, A1.

38. Susan Braedley and Meg Luxton, eds., *Neoliberalism and Everyday Life* (Montreal: McGill-Queens University Press, 2010). See also Lisa Duggan, *The Twilight of Equality? Neoliberalism, Cultural Politics, and the Attack on Democracy* (Boston: Beacon, 2004).

39. Bellah, Madsen, Sullivan, Swidler, and Tipton, *Habits of the Heart*, 127.

40. McMahon, "Do You Have a Panglossian Disorder?"

41. Carl Boggs, *The End of Politics: Corporate Power and the Decline of the Public Sphere* (New York: Guilford, 2000), 178.

42. McMahon, "Do You Have a Panglossian Disorder?"

43. "Voters Issue Strong Rebuke of Incumbents in Congress," Gallup, April 7, 2010, http://www.gallup.com/poll/127241/voters-issue-strong-rebuke -incumbents-congress.aspx.

44. Carl Boggs, "Warrior Nightmares: American Reactionary Populism at the Millennium," *Socialist Register* 36 (2000): 243–56, 247.

45. Richard J. Mitchell Jr., *Dancing at Armageddon: Survivalism and Chaos in Modern Times* (Chicago: University of Chicago Press, 2002), 11, 10.

46. Respondent 9707082.

47. Sixty-four percent of American respondents who described themselves as liberal or very liberal had stockpiled food, compared to 78 percent of respondents who were conservative or very conservative.

48. See Jeffrey Jacob, *New Pioneers: The Back-to-the-Land Movement and the Search for a Sustainable Future* (University Park: Pennsylvania State University Press, 1997).

49. Raymond J. Coffin and Mark W. Lipsey, "Moving Back to the Land: An Ecologically Responsible Lifestyle Change," *Environment and Behavior* 13, no. 1 (1981): 42–63, 57. Most back-to-the-landers had additional motivations for moving, of course—88 percent said that they wanted "to get closer to nature," while 83 percent wanted the "chance to grow and eat wholesome food."

50. Jacob, *New Pioneers*, 3.

51. Thierry Bardini, *Bootstrapping: Douglas Engelbart, Coevolution, and the Origins of Personal Computing* (Stanford, CA: Stanford University Press, 2000), 216.

52. Janet Abbate, *Inventing the Internet* (Cambridge, MA: MIT Press, 1999), 8–13.

53. Jeanette Hofmann, "The Libertarian Origins of Cybercrime: Unintended Side-Effects of a Political Utopia," Discussion Paper no. 62 (London: Centre for Analysis of Risk and Regulation, April 2010).

54. Jack Goldsmith and Tim Wu, *Who Controls the Internet? Illusions of a Borderless World* (Oxford: Oxford University Press, 2006), 23.

55. Fred Turner, *From Counterculture to Cyberculture: Stewart Brand, the Whole Earth Network, and the Rise of Digital Utopianism* (Chicago: University of Chicago Press, 2006).

56. Philip Jenkins, *Decade of Nightmares: The End of the Sixties and the Making of Eighties America* (New York: Oxford University Press, 2006), 193.

57. See, e.g., the films *WarGames*, John Badham (1983), *Prime Risk*, Michael L. Farkas (1985), *Terminal Entry*, John Kincade (1986), *Ferris Bueller's Day Off*, John Hughes (1986), and, in the early 1990s, *Sneakers*, Phil Alden Robinson (1992), and *Hackers*, Iain Softley (1995).

58. Thomas Streeter, *The Net Effect: Romanticism, Capitalism, and the Internet* (New York: New York University Press, 2010), 88.

59. John Bellamy Foster and Robert W. McChesney, "The Internet's Unholy Marriage to Capitalism," *Monthly Review: An Independent Socialist Magazine* 62, no. 10 (2011), http://monthlyreview.org/2011/03/01/the-internets -unholy-marriage-to-capitalism.

60. William J. Clinton and Albert Gore Jr., "A Framework for Global Electronic Commerce," July 1, 1997, http://clinton4.nara.gov/WH/New/Commerce/ read.html.

61. Paulina Borsook, *Cyberselfish: A Critical Romp through the Terribly Libertarian Culture of High Tech* (New York: PublicAffairs, 2000), 130.

62. Bill Gates, *The Road Ahead* (New York: Penguin, 1995), 7; Christopher Anderson, "The Accidental Superhighway," *Economist*, July 1, 1995, 5–20, 4. Anderson would later become the editor-in-chief of *Wired*.

63. Foster and McChesney, "The Internet's Unholy Marriage to Capitalism."

64. Turner, *From Counterculture to Cyberculture*, 215.

65. Sherry Turkle, *Life on the Screen: Identity in the Age of the Internet* (New York: Simon & Schuster, 1997); Putnam, *Bowling Alone*; Robert Kraut, Michael Patterson, Vicki Lundmark, Sara Kiesler, Tridas Mukophadhya, and William Scherlis, "Internet Paradox: A Social Technology That Reduces Social Involvement and Psychological Well-Being?" *American Psychologist* 53, no. 9 (1998): 1017–31. See also N. H. Nie and L. Erbring, *Internet and Society: A Preliminary Report* (Stanford, CA: Institute for Quantitative Studies of Society, 2000).

66. Shelley Boulianne, "Does Internet Use Affect Engagement? A Meta-Analysis of Research," *Political Communication* 26, no. 2 (2009): 193–211.

67. Nicholas Carr, *The Shallows: What the Internet Is Doing to Our Brains* (New York: Norton, 2010), 2.

68. Robert Hassan, "Social Acceleration and the Network Effect: A Defence of Social 'Science Fiction' and Network Determinism," *British Journal of Sociology* 61, no. 2 (2010): 356–74.

69. The social construction approach emerged in the mid- to late 1980s with the publication of Wiebe E. Bijker, Thomas P. Hughes, and Trevor J. Pinch, eds., *The Social Construction of Technological Systems: New Directions in the Sociology and History of Technology* (Cambridge, MA: MIT Press, 1987); and Donald MacKenzie and Judy Wajcman, eds., *The Social Shaping of Technology* (Milton Keynes: Open University Press, 1985).

70. Hassan, "Social Acceleration and the Network Effect," 364.

71. Ibid., 367. The use of a universalizing *we* in this quote seems to ignore the digital divide, the significant class- and race-based gaps in access to technologies, even within Western industrialized nations.

72. Langdon Winner, "Technology as Forms of Life," in *Readings in the Philosophy of Technology*, ed. David M. Kaplan (Oxford: Rowman & Littlefield, 2004), 103–13.

73. One suspects that the balkanization of academic fields and disciplines, combined with the increase in authors and scholarship, is also partially responsible here. Both trends have encouraged a focus on smaller and smaller phenomena. As a result, popular nonfiction authors, willing to make more ambitious claims at the risk of overgeneralizing, have at times produced more insightful analyses of the network as a general phenomenon.

74. Manuel Castells, *The Rise of the Network Society* (Cambridge, MA: Blackwell, 1996), 2.

75. Evgeny Morozov, *The Net Delusion: The Dark Side of Internet Freedom* (New York: PublicAffairs, 2011), 295.

76. Karel Kreijns and Paul A. Kirschner, "The Social Affordances of Computer-Supported Collaborative Learning Environments" (paper presented at the ASEE/IEEE Frontiers in Education Conference, Reno, NV, October 2001).

77. Wellman, "Little Boxes, Glocalization, and Networked Individualism," 338.

78. Darin Barney, *The Network Society* (Malden, MA: Polity, 2004), 172.

79. Ima Tubella, "Television and Internet in the Construction of Identity," in *The Network Society: From Knowledge ot Policy*, ed. Manuel Castells (Washington, DC: Johns Hopkins Center for Transatlantic Relations, 2008), 257–68, 258.

80. Respondents 9684580, 9576666, 9576729, 9560116.

81. Catherine M. Ridings and David Geffer, "Virtual Community Attraction: Why People Hang Out Online," *Journal of Computer-Mediated Communication* 10, no. 1 (2004), http://onlinelibrary.wiley.com/doi/10.1111/j.1083-6101.2004.tb00229.x/abstract.

82. Interview with Kris Can, Denver, CO, November 12, 2009.
83. Although my two surveys, which were conducted online, will necessarily show a bias toward peakists who at least have Internet access, interviews that were conducted in person showed a similar distribution of participation in peak oil Web sites.
84. The Oil Age, http://www.theoilage.com (now defunct); Collapse Network, http://www.collapsenet.com; The Archdruid Report, http://thearchdruidreport.blogspot.com; Peak Oil Crisis, http://www.peak-oil-crisis.com; Decline of the Empire, http://www.declineoftheempire.com; The Post-Carbon Institute, http://www.postcarbon.org; Peak Moment Television, http://www.peakmoment.tv; Malthusia, http://www.malthusia.com; Silent Country, http://forums.silentcountry.com/forums; Early Warning, http://earlywarn.blogspot.com.
85. In response to the question, "Have you experienced any of the following as a result of your awareness of peak oil?" 31.7 percent of respondents answered "Gained friends." However, as we saw in chapter 1, many previous relationships were strained by the conversion to peakism—21.6 percent answered "Lost friends."
86. "Don't Do It," Peak Oil News, n.d., http://peakoil.com/forums/viewtopic.php?f=35&t=49470&view=unread (post deleted).
87. "My Doom-O-Meter is jittering toward max," Peak Oil News, March 11, 2008, http://peakoil.com/forums/my-doom-o-meter-is-jittering-towards-max-t37601-15.html#p615813.
88. Respondents 9569428, 9680437.
89. See, e.g., Gustavo Mesch and Ilan Talmud, "The Quality of Online and Offline Relationships: The Role of Multiplexity and Duration of Social Relationships," *Information Society: An International Journal* 22, no. 3 (2006): 137–48; and Ines Mergel and Thomas Langenberg, "What Makes Online Ties Sustainable? A Research Design Proposal to Analyze Online Social Networks," Working Paper no. PNG06-002 (Cambridge, MA: Program on Networked Governance, John F. Kennedy School of Government, Harvard University, 2006).
90. Mark S. Granovetter, "The Strength of Weak Ties," *American Journal of Sociology* 78 (1973): 1360–80, 1361.
91. "I'd Like to Meet All of You," Peak Oil News, January 28, 2008, http://peakoil.com/forums/i-d-love-to-meet-all-of-you-t36002.html.
92. See Miller McPherson, Lynn Smith-Lovin, and Matthew E. Brashears, "Social Isolation in America: Changes in Core Discussion Networks over Two Decades," *American Sociological Review* 71, no. 3 (2006): 353–75.
93. Rodgers, *The Age of Fracture*, 220.
94. See, e.g., Nicole B. Ellison, Charles Steinfield, and Cliff Lampe, "The Benefits of Facebook 'Friends': Social Capital and College Students' Use of Online Social Network Sites," *Journal of Computer-Mediated Communication*

12, no. 4 (2007): 1143–68. This point primarily refers to early analyses of Facebook since some individuals now use the platform to make connections with people they have met only online.

95. "Re: What Happened to the Peak Oil Movement?" Peak Oil News, February 10, 2011, http://peakoil.com/forums/what-happened-to-the-peak-oil-movement-t60782-30.html.

96. John Drury, Christopher Cocking, Joseph Beale, Charlotte Hanson, and Faye Rapley, "The Phenomenology of Empowerment in Collective Action," British Journal of Social Psychology 44, no. 3 (2005): 309–28, 317, 320.

97. Tom Postmes, "The Psychological Dimensions of Collective Action, Online," in The Oxford Handbook of Internet Psychology, ed. Adam M. Joinson (London: Oxford University Press, 2007), 165–86, 166.

98. Blair Nonnecke and Jenny Preece, "Lurker Demographics: Counting the Silent" (paper presented at the Conference on Human Factors in Computing Systems, The Hague, Netherlands, April 2000).

99. Respondents 9570685, 9601959.

100. John Drury and Steve Reicher, "Collective Psychological Empowerment as a Model of Social Change: Researching Crowds and Power," Journal of Social Issues 65, no. 4 (2009): 707–25, 720.

101. Albert Bandura, "Exercise of Human Agency through Collective Efficacy," Current Directions in Psychological Science 9, no. 3 (2000): 75–78.

102. Postmes, "The Psychological Dimensions of Collective Action, Online," 166.

103. Robert Nozick, Anarchy, State and Utopia (New York: Basic, 1974), 33; Thatcher quoted in Jeff Popke, "Latino Migration and Neoliberalism in the U.S. South: Notes toward a Rural Cosmopolitanism," Southeastern Geographer 51, no. 2 (2011): 242–59, 252.

104. Henry A. Giroux, "Democracy, Patriotism, and Schooling After September 11th: Critical Citizens or Unthinking Patriots?" in The Abandoned Generation: Democracy beyond the Culture of Fear (New York: Macmillan, 2003), 16–45, 31.

105. Insofar as part of the motivation for the decentralized Internet was the threat of a previous environmental (nuclear) apocalypse and the influence of libertarian environmentalists such as Stewart Brand, we might see the default libertarianism of peakists as one moment in a longer dialectic between socialization and collectivism.

CHAPTER FOUR

1. Daniel Wojcik, The End of the World as We Know It: Faith, Fatalism and Apocalypse in America (New York: New York University Press, 1999), 12.

2. "Y2K Millennium Bug—Wave 3," Gallup (poll sponsored by USA Today/ National Science Foundation), August 25–29, 1999, http://brain.gallup

.com/documents/questionnaire.aspx?STUDY=P9908039&p=2; Pew
Research Center, "Public Sees a Future Full of Promise and Peril: Life in
2050: Amazing Science, Familiar Threats," June 22, 2010, http://www
.peoplepress.org/2010/06/22/section-3-war-terrorism-and-global-trends.

3. Nancy Gibbs, "Apocalypse Now," Time, July 1, 2002, 38–46.

4. Seventy-two percent of American respondents to my surveys had "prepared food or other supplies for yourself and your family" as "a result of
your knowledge of peak oil," compared to 60 percent of Canadians and
Western Europeans.

5. See, e.g., Ernest Lee Tuveson, Redeemer Nation: The Idea of America's Millennial Role (Chicago: University of Chicago Press, 1968).

6. Mark Hulsether, Religion, Culture and Politics in the Twentieth-Century United
States (New York: Columbia University Press, 2007), 138–71.

7. Michael S. Northcott, An Angel Directs the Storm: Apocalyptic Religion and
American Empire (London: SCM, 2007).

8. See, e.g., James H. Moorhead, American Apocalypse: Yankee Protestants and
the Civil War, 1860–1869 (New Haven, CT: Yale University Press, 1978);
Merton L. Dillon, "The Failure of the American Abolitionists," Journal of
Southern History 25, no. 2 (1959): 159–77; Joel W. Martin, "Before and Beyond the Sioux Ghost Dance: Native American Prophetic Movements and
the Study of Religion," Journal of the American Academy of Religion 59, no. 4
(1991): 677–701; and Wojcik, The End of the World as We Know It, chap. 2.

9. See, e.g., Vaclav Smil, "Peak Oil: A Catastrophic Cult and Complex Realities," World Watch Magazine 19, no. 1 (2006): 22–24; Graham Chandler,
"The Cult of Peak Oil," Alberta Oil: The Business of Energy, October 1, 2007,
http://www.albertaoilmagazine.com/2007/10/the-cult-of-peak-oil; and
Urstadt, "Imagine There's No Oil."

10. See Taylor, "Exploring Religion, Nature and Culture."

11. See, e.g., the spirited debate in a thread on Peak Oil News and Message
Boards called "Peak oil is a religion," http://peakoil.com/forums/peak-oil
-is-a-religion-t52427.html.

12. See, e.g., Rome, "'Give Earth a Chance'"; Paul Boyer, By the Bomb's Early
Light (New York: Pantheon, 1985); and Elaine Tyler May, Homeward Bound:
American Families in the Cold War Era (New York: Basic, 1990).

13. Michael Barkun, "Millennium Culture: The Year 2000 as a Religious
Event," in Millennial Visions: Essays on Twentieth-Century Millenarianism,
ed. Martha F. Lee (Westport, CT: Praeger, 2000), 41–54.

14. Kevin Sack, "Apocalyptic Theology Revived by Attacks," New York Times,
November 23, 2001, A17.

15. Mervyn F. Bendle, "The Apocalyptic Imagination and Popular Culture,"
Journal of Religion and Popular Culture 11 (Fall 2005): 1–42.

16. Douglas Kellner, Cinema Wars: Hollywood Film and Politics in the Bush-
Cheney Era (Oxford: Wiley-Blackwell, 2010), 123. Of the sixteen Left Behind

novels, written by Tim Lahaye and Jerry B. Jenkins, seven have reached number 1 on the *New York Times*, *USA Today*, or *Publishers Weekly* bestseller lists, and the series has sold over sixty-five million books.

17. My definition of *disaster films* follows Nick Roddick's 1980 influential definition. They have the following characteristics: the disasters must be central to the film as well as "factually possible," "largely indiscriminate," and "unexpected (though not necessarily unpredictable)." Nick Roddick, "Only the Stars Survive: Disaster Movies in the Seventies," in *Performance and Politics in Popular Drama: Aspects of Popular Entertainment in Theatre, Film and Television, 1800–1976*, ed. David Bradby, Louis James, and Bernard Sharratt (Cambridge: Cambridge University Press, 1980), 243–70, 246.

18. A full list of disaster films of the period 2000–2010 includes *The Core* ($73 million), *The Day After Tomorrow* ($187 million), *I, Robot* ($144 million), *War of the Worlds* ($234 million), *World Trade Center* ($163 million), *Children of Men* ($35 million), *Poseidon* ($60 million), *I Am Legend* ($256 million), *Sunshine* ($32 million), *Cloverfield* ($80 million), *The Day the Earth Stood Still* ($79 million), *The Happening* ($31 million), *Doomsday* ($11 million), *2012* ($166 million), *Knowing* ($80 million), *Battle: Los Angeles* ($212 million), and *Contagion* ($135 million). This list does not include *Titanic* ($600 million), which has a much more limited scope than other disaster films of the period, or films like *The Matrix* and *Fight Club*, which carry apocalyptic themes but would not be classified as disaster movies. The genre is even more popular among made-for-television films—from 2000 to 2011, over fifty disaster movies were released on American television networks.

19. "I Always Stop If *Twister* or *Independence Day* Is on TV," http://www .facebook.com/pages/I-Always-Stop-if-Twister-or-Independence-Day-Is-on -Tv/113417768671144.

20. Geoff King, "'Just Like a Movie'? 9/11 and Hollywood Spectacle," in *The Spectacle of the Real: From Hollywood to "Reality" TV and Beyond*, ed. Geoff King (Portland, OR: Intellect, 2005), 47–57, 48; Neal Gabler, "This Time, the Scene Was Real," *New York Times*, September 16, 2001, sec. 4, p. A2. The Motion Picture Association of America also took note of the connection between disaster films and 9/11 and began to consider "disaster images" as a new criterion for applying a PG-13 rating.

21. Benjamin Svetkey, "Lava Is a Many Splendored Thing," *Entertainment Weekly*, August 25, 1997, 3.

22. Ryan Gilbey, "Climate Change Is Inspiring the Ultimate Scary Movies," *Guardian*, January 1, 2010, 28.

23. Stephen Keane, *Disaster Movies: The Cinema of Catastrophe* (London: Wallflower, 2001), 17.

24. See, e.g., J. Hoberman, "*Nashville* contra *Jaws*; or, The Imagination of Disaster Revisited," in *The Last Great American Picture Show: New Hollywood Cinema in the 1970s*, ed. Alexander Horwath, Thomas Elsaesser, and Noel

King (Amsterdam: Amsterdam University Press, 2004), 195–222; and David A. Cook, *Lost Illusions: American Cinema in the Shadow of Watergate and Vietnam, 1970–1979* (New York: Scribner, 1999).

25. Keane, *Disaster Movies*, 101.

26. See, e.g., *Independence Day* (1996), *Deep Impact* (1998), and *The Day After Tomorrow* (2004).

27. David Denby, "Lava Story," *New York*, May 12, 1997, 58. See also Ken Feil, *Dying for a Laugh: Disaster Movies and the Camp Imagination* (Middletown, CT: Wesleyan University Press, 2006), 134.

28. Feil, *Dying for a Laugh*, 1.

29. Keane, *Disaster Movies*, 63.

30. If any demonstration of this point is necessary, see Michelle Alexander, *The New Jim Crow: Mass Incarceration in the Age of Colorblindness* (New York: New Press, 2010).

31. Despina Kakoudaki, "Spectacles of History: Race Relations, Melodrama, and the Science Fiction/Disaster Film," *Camera Obscura* 17, no. 2 (2002): 109–53, 113.

32. While I have focused here on disaster narratives in film, these elements hold true for the vast majority of secular disaster narratives in popular culture, such as television shows, video games, and the wide range of films that feature epic scenes of destruction but are not categorized as disaster movies per se. Despite their differences, what these media have in common is a narrative that ultimately views a potentially apocalyptic crisis as a means of national regeneration. Of course, this does not hold true for all recent films that contemplate the end of the world or feature spectacles of destruction. *The Road*, both the film and novel, is a prominent example— although it ends on a note of relative hope, it offers a bleak, depressing, and unredemptive vision of the future. However, I would argue that *The Road*, like *On the Beach* (also a novel and a film) before it, is jarring because its pessimistic prophecy goes so strongly against the generic grain. Most religious apocalyptic narratives, especially Christian (particularly evangelical) stories such as the *Left Behind* series, hold the expectation that believers (the reader) will enjoy the millennium of Christ's reign after the apocalypse.

33. Stuart Hall, "Encoding and Decoding in the Television Discourse" (paper presented to the Council of Europe Colloquy, University of Leicester, September 1973).

34. See, e.g., Robert Willson, "*Jaws* as Submarine Movie," *Jump Cut: A Review of Contemporary Media* 15 (1977): 32–33; Jane Caputi, "*Jaws* as Patriarchal Myth," *Journal of Popular Film* 6, no. 4 (1978): 305–26; and Frederic Jameson, "Reification and Utopia in Mass Culture," *Social Text* 1 (1979): 130–48.

35. A number of pop cultural texts that were also influential to peak oil believers, including the film *Mad Max* (1979) and apocalyptic novels like

Larry Niven and Jerry Pournelle's *Lucifer's Hammer* (Santa Monica, CA: Playboy Press, 1977), are not discussed in this chapter. Rather than explore all the pop cultural (and media) influences on peak oil believers, this chapter looks at one iteration of American apocalyptic pop culture, but it is not intended to deny the influence of others.

36. Respondent 9589640.
37. In the study of media effects, priming is a process whereby viewers of media become more likely to think about a certain topic in a certain way owing to continued exposure. In this context, it is the quantity of disaster films that is most relevant. See Glenn G. Sparks, *Media Effects Research* (Boston: Wadsworth, 2010), 105–6, 200–201.
38. Respondents 9570329, 9592954, 9575686.
39. Respondents 9685851, 9686672, 9686672, 9680234.
40. Balmford et al., "Hollywood, Climate Change, and the Public," 1713. In this study, concern was measured by the audience's allocation of hypothetical money to climate change mitigation (as opposed to four other worthy causes) and the emissions-reducing actions that respondents said they planned to take before and after seeing the film.
41. Ruppert has also published *Confronting Collapse: The Crisis of Energy and Money in a Post Peak Oil World* (White River Junction, VT: Chelsea Green, 2009) and appeared in a number of peak oil films, such as *The End of Suburbia*.
42. "Nostra," "My Movie Influence: *Collapse* (2009)," Inspired Ground, April 16, 2012, http://www.inspired-ground.com/my-movie-influence -collapse-2009.
43. "SnakelaD," "The Movie That Changed My Life," Amazon.com, n.d., http://www.amazon.com/Collapse-Michael-Ruppert/product-reviews/ B003CJXJ8Q?pageNumber=7 (post now removed).
44. Respondents 9591539, 9591884.
45. Respondents 9566782, 9695012, 9567344, 9584720, 9685028, 9595099, 9689841, 9685028.
46. In response to the question, "Given the peak oil crisis, which of the following do you see as most likely for the country you currently reside in?" American peakists averaged 6.4 of 10 for the option "apocalyptic scenario (i.e., violence, epidemics, die-off)," while Canadians and Western Europeans averaged 5.6. There are many potential explanations for this difference, of course, but the long tradition of American apocalypticism is certainly relevant and was frequently cited by respondents themselves.
47. Respondents 9568275, 9568924.
48. As of March 3, 2012, the "Planning for the Future Thread" of Peak Oil News contained 2,695 topics. The most popular, "Today I made / bought / learn . . . (for a post oil world)," contained 1,936 posts and had been viewed 305,880 times. "Planning for the Future," Peak Oil News and Message Boards, http://peakoil.com/forums/planning-for-the-future-f8.html.

49. Wojcik, *The End of the World as We Know It*, 16.

50. Amitav Ghosh, "Petrofiction: The Oil Encounter and the Novel," *New Republic*, March 2, 1992, 29–34. While this lacuna was particularly glaring in literature, American artists and purveyors of popular culture have tended to ignore energy except when it is new, controversial, or expensive. Recent years have seen a flowering of oil culture, by artists such as Marina Zurkow and Edward Burtynsky. On the academic side, this has been reflected by an increased awareness of the relationship between cultural expression and energy. See, e.g., LeMenager, *Living Oil*; and Szeman, Wenzel, and Yaeger, eds., *Fueling Culture*.

51. Global oil consumption from 1992 to 2012 calculated using figures from the US Energy Information Administration, Independent Statistics and Analysis, "International Petroleum (Oil) Consumption Tables (Thousand Barrels per Day)," n.d., http://www.eia.gov/emeu/international/oilconsumption.html.

52. Jim Booth, *Boil Over: The Day the Oil Ran Out* (Charleston, SC: CreateSpace, 2010); John M. Cape and Laura Buckner, *Oil Dusk: A Peak Oil Story* (Humble, TX: Singing Bowl, 2009); Kurt Cobb, *Prelude* (Pittsburgh: Public Interest Communications, 2010); Douglas Coupland, *Player One: What Is to Become of Us* (Toronto: House of Anansi, 2010); William Flynn, *Shut Down: A Story of Economic Collapse and Hope* (Charleston, SC: CreateSpace, 2011); Johnston, *After the Crash*; James Howard Kunstler, *World Made by Hand* (New York: Atlantic Monthly Press, 2008), and *The Witch of Hebron* (New York: Atlantic Monthly Press, 2010); Sam Penny, *Was a Time When* (Sacramento, CA: TwoPenny, 2012); Amy Rogers, *Petroplague* (New York: Diversion, 2011); Alex Scarrow, *Last Light* (London: Orion, 2007), and *Afterlight* (London: Orion, 2010); Dave Spicer, *Deadly Freedom* (New York: Cambridge House, 2008). For an example of the marketing of these novels within the peakist community, see Caryl Johnston, "First Peak Oil Novel Published in U.S.!" *Energy Bulletin*, August 15, 2005, http://www.energybulletin.net/node/7921.

53. See, e.g., Frank Kaminski, "The Post-Oil Novel: A Celebration!" Seattle Peak Oil Awareness, May 11, 2008, http://www.energybulletin.net/node/44031.

54. Cobb, *Prelude*, 153, 147.

55. I rely here on the hundreds of reviews and virtual discussions of *Last Light* on peak oil Web sites and blogs. See also Brendan Barrett, "Reality of Peak Oil Enters Our Fiction," OurWorld 2.0, January 21, 2011, http://ourworld.unu.edu/en/reality-of-peak-oil-enters-our-fiction.

56. Scarrow, *Last Light*, 398.

57. For two of the few critics who have highlighted this aspect of apocalyptic texts, see George Slusser, "Pocket Apocalyse: American Survivalist Fictions from *Walden* to *The Incredible Shrinking Man*," in *Imagining Apocalypse: Studies in Cultural Crisis*, ed. David Seed (New York: St. Martin's, 2000), 118–35;

and Claire P. Curtis, *Postapocalyptic Fiction and the Social Contract: "We'll Not Go Home Again"* (Lanham, MD: Rowman & Littlefield, 2012), 5.

58. Scarrow, *Last Light*, 402 (emphasis added).

59. Respondents 9682436, 9573106.

60. Psychologists and sociologists have noted that this tendency explains why many apocalypticists double down on their belief when predicted events do not come to pass. See, e.g., the classic study of failed predictions, Leon Festinger's *When Prophecy Fails* (Minneapolis: University of Minnesota Press, 1956).

61. "How Many People Here Are Rooting for Peak Oil?" Peak Oil News, May 29, 2006, http://peakoil.com/forums/how-many-people-here-are -rooting-for-peak-oil-t20704.html.

62. Respondents 9559210, 9561426, 9561825, 9563057. On this subject, Daniel Wojcik noted: "The appeal of apocalypticism may also be attributable to the fact that such beliefs enhance the self-esteem of believers. Revelation of the secret order of events in the midst of seeming chaos makes devotees privy to arcane knowledge of the meaning of history." Wojcik, *The End of the World as We Know It*, 143–44.

63. Paul Greenberg, "Recipes for Disaster," *New York Times*, April 20, 2008, 27; McGrath, "The Dystopians"; Rebecca Onion, "Envirogeddon! Is It Time to Start Wishing for the End of the World?" Slate, April 21, 2008, http://www .unz.org/Pub/Slate-2008apr-00291; *The Colbert Report*, Comedy Central, May 1, 2008.

64. Respondent 9581634.

65. Kunstler, *World Made by Hand*, 220.

66. Ibid., 76, 5, 148.

67. "Polling the Tea Party," *New York Times*/CBS News poll, April 5–12, 2010. This poll was used to measure the "racial attitudes" of Tea Party support- ers, who, while being very similar to peakists in terms of race, age, educa- tion, and wealth, are diametrically opposed on this and other political issues. In this poll, e.g., 52 percent of Tea Party supporters answered that "too much" had "been made of the problems facing black people." View the data at http://www.nytimes.com/interactive/2010/04/14/us/politics/ 20100414-tea-party-poll-graphic.html?ref=politics#tab=5. William Luther Pierce is the author of *The Turner Diaries* (Hillsboro, WV: National Vanguard, 1978), an anti-Semitic post-apocalyptic novel that was an inspi- ration for the racist Right and inspired quasi-survivalists such as Timothy McVeigh.

68. Matthew T. Huber, "The Use of Gasoline: Value, Oil, and the 'American Way of Life,'" *Antipode* 41, no. 3 (2009): 465–86, 469, 476.

69. As Imre Szeman put it, the filmic technique of presenting "multiple story- lines which take place in numerous locations" is "a now common, overly literal attempt to represent the new reality of globalization" and the inter-

connectedness of different nations and regions. Szeman, "The Cultural Politics of Oil," 36.

70. Benedict Anderson, *Imagined Communities: Reflections on the Spread of Nationalism* (London: Verso, 1983); Kunstler, *World Made by Hand*, 317.

71. James L. Howard, "Global Peak Oil Survey 2009," September 27, 2009, http://www.powerswitch.org.uk/portal/index.php?option=com_content &task=view&id=2971&Itemid=77. This was an online survey of nearly three hundred peakists. While Howard's and my survey respondents could be slightly different considered as a population, his survey's demographics (on age, gender, country of origin, and a number of other factors) indicate that it is a nearly perfect match with mine.

72. Kunstler, *World Made by Hand*, 38.

73. Reihan Salam, "Heralding the End Times," *New York Sun*, March 5, 2008, Arts and Letters section, 15.

74. Leo Marx, *The Machine in the Garden: Technology and the Pastoral Ideal in America* (New York: Oxford University Press, 1964), 5.

75. Henry Nash Smith, *Virgin Land: The American West as Symbol and Myth* (Cambridge, MA: Harvard University Press, 1950), 170, 126.

76. Kunstler, *World Made by Hand*, 32, 142, 58.

77. The environmental perspective of this novel, as is true of many pastoral fantasies, is reliant on a socially constructed and often anthropocentric conception of what is good or bad for this thing we call *the environment*. For a history of the social construction of nature, see Donald Worster, *Nature's Economy: A History of Ecological Ideas* (New York: Cambridge University Press, 1985).

78. Kunstler, *World Made by Hand*, 5, 309.

79. The historiography on crises of masculinity in the United States is long and varied. See, e.g., John F. Kasson, *Houdini, Tarzan, and the Perfect Man: The White Male Body and the Challenge of Modernity in America* (New York: MacMillan, 2002).

80. See Curtis, *Postapocalyptic Fiction and the Social Contract*, chap. 1; Peter Hitchcock, "Oil in an American Imaginary," *New Formations* 69, no. 1 (2010): 81–97; Stephanie LeMenager, "The Aesthetics of Petroleum, After Oil!" *American Literary History* 24, no. 1 (2012): 59–86; and Mary Manjikian, *Apocalypse and Post-Politics: The Romance of the End* (Lanham, MD: Rowman & Littlefield, 2012).

81. Lydia Saad, "Conservatives Remain the Largest Ideological Group in U.S.," Gallup, January 12, 2012, http://www.gallup.com/poll/152021/ conservatives-remain-largest-ideological-group.aspx.

82. In a *Washington Post*–ABC poll taken two months before this survey, 53 percent of Americans agreed that same-sex marriage should be legal, with 44 percent against. Sandhya Somashekhar and Peyton Craighill, "Slim Majority Back Gay Marriage, *Post*-ABC Poll Says," *Washington Post*, March 18, 2011, http://articles.washingtonpost.com/2011-03-18/politics/

35207422_1_gay-marriage-pro-gay-marriage-group-anti-gay-marriage. The polling statement, "We don't give everyone an equal chance in this country," has been used by researchers to measure sensitivity to (and resentment of) minority groups. In a 2010 survey, the University of Washington Institute for the Study of Ethnicity, Race and Sexuality found that only 55 percent of white Americans agreed with this statement. Christopher Parker, "Attitudes on Limits to Liberty, Equality, and Pres. Obama Traits by Tea Party Approval," University of Washington Institute for the Study of Ethnicity, Race and Sexuality, 2010, http://depts.washington.edu/uwiser/Tea%20Party%20Chart%20%5Bpdf%5D-1.pdf.

83. "The American Dream?" Peak Oil News, April 19, 2011, http://peakoil .com/forums/the-american-dream-t61397.html; "16 Reasons to Feel Really Depressed about the Direction That the Economy Is Heading," Peak Oil News, July 7, 2011, http://peakoil.com/business/16-reasons-to-feel-really -depressed-about-the-direction-that-the-economy-is-headed.

84. "Breaking: US AAA Credit Rating Downgraded," Peak Oil News, August 6, 2011, http://peakoil.com/forums/breaking-us-aaa-credit-rating -downgraded-t62368-105.html.

85. Respondents 9589678, 9697109, 9589863, 9571401, 9590289.

86. "How Reliable Are U.S. Department of Energy Oil Production Forecasts?" Peak Oil News, February 17, 2012, http://peakoil.com/production/how -reliable-are-u-s-department-of-energy-oil-production-forecasts.

87. Imre Szeman, "System Failure: Oil, Futurity, and the Anticipation of Disaster," *South Atlantic Quarterly* 106, no. 4 (2007): 805–23, 808.

88. See, e.g., Octavia Butler, *Parable of the Sower* (New York: Four Walls Eight Windows, 1993); and Ursula K. LeGuin, *The Dispossessed: An Ambiguous Utopia* (New York: Harper & Row, 1974).

89. Mike Bendzela, "The End of the End: How the Peak Oil Movement Failed," OpEdNews, January 2, 2011, http://www.opednews.com/articles/The-End -of-The-End—How-t-by-Mike-Bendzela-110102-802.html.

90. In response to the question, "Have you been more or less engaged in the following activities since learning about peak oil? Please put on a scale from 1 to 5, with 1 for much less engaged, 3 for no change, and 5 for much less engaged," American peakists averaged 2.79 for the option "attending rallies, marches or protests."

91. Respondents 9567344, 9591372, 9591897.

92. Respondent 9596038.

93. "Oily Cassandra," personal communication, February 1, 2010.

94. "'We Were Lied To,'" Clusterfuck Nation, June 2, 2008, http://kunstler .com/mags_diary24.html.

95. McMahon, "Do You Have a Panglossian Disorder?"

96. "Voters Issue Strong Rebuke of Incumbents in Congress." Some survey questions mirrored, word for word, national polls conducted regularly over the last thirty years.

97. Brian Montopoli, "Alienated Nation: Americans Complain of Government Disconnect," CBS News, June 28, 2011, http://www.cbsnews.com/news/alienated-nation-americans-complain-of-government-disconnect.

98. Timothy Weber, *Living in the Shadow of the Second Coming: American Premillennialism, 1875–1982* (New York: Oxford University Press, 1979), x.

99. "Harris Poll Alienation Index Climbs Again as Two-Thirds of Americans Feel Alienated," Harris Polls, November 12, 2013, http://www.harrisinteractive.com/NewsRoom/HarrisPolls/tabid/447/ctl/ReadCustom%20Default/mid/1508/ArticleId/1316/Default.aspx.

100. Rick Couri, "Poll Places Trust in Congress below Telemarketers," Newstalk Radio KRMG, December 13, 2011, http://www.krmg.com/news/news/local/poll-places-trust-congress-below-telemarketers/nFz8m.

101. Over the last forty years, voter turnout in the United States has hovered between 36 and 57 percent, with nonpresidential elections generally in the 30s and 40s. This is far below other industrialized countries. For lower house elections, the US average for the eighteen elections before 2001 was 48 percent turnout of eligible voters. Compare this rate of participation over the same period to those in Germany (86 percent), Holland (83 percent), Israel (80 percent), France (76 percent), the United Kingdom (76 percent), Canada (74 percent), and Spain (73 percent). See Mark N. Franklin, "Electoral Participation," in *Controversies in Voting Behavior*, ed. Richard G. Niemi and Herbert F. Weisberg (Washington, DC: CQ Press, 2001), 83–99.

102. Respondent 9686274.

103. "Carlhole," "Doomers Don't Believe in America: Economics and Finance—Page 9," Peak Oil News, September 25, 2008, http://peakoil.com/forums/doomers-don-t-believe-in-america-t45511-120.html.

104. As Bill McKibben noted in a recent *Rolling Stone* article, the value of all the announced reserves of coal, oil, and natural gas, held by countries and corporations, is $27 trillion. McKibben, "Global Warming's Terrifying New Math."

105. See, e.g., Eli Kintisch, *Hack the Planet: Science's Best Hope—or Worst Nightmare—for Averting Climate Catastrophe* (Hoboken, NJ: Wiley, 2011); and Clive Hamilton, *Earthmasters: The Dawn of the Age of Climate Engineering* (New Haven, CT: Yale University Press, 2013).

106. Of course, as scholars such as James Scott have shown (e.g., James C. Scott, *Weapons of the Weak: Everyday Forms of Peasant Resistance* [New Haven, CT: Yale University Press, 1985], and *Domination and the Arts of Resistance: Hidden Transcripts* [New Haven, CT: Yale University Press, 1990]), there are many kinds of more and less visible political activity, of which participation in formal electoral politics and traditional political dissent are only the most public. This is certainly true, and this analysis is intended to assert not that only visible political activities are valuable but only that they are the most direct means of achieving the kinds of broad changes that peakists are interested in. Whether the retreat of some peakists from

formal, visible political activities is accompanied by an increase in other activities that might be termed *political* is an important question, but one that is slightly beyond the methodology of this book.

107. See, e.g., Michael Hardt, "Two Faces of Apocalypse: A Letter from Copenhagen," *Polygraph* 22 (2010): 265–74.

108. Constance Penley, "Time Travel, Primal Scene, and the Critical Dystopia," in *Close Encounters: Film, Feminism, and Science Fiction*, ed. Constance Penley, Elizabeth Lyon, Lynn Spigel, and Janet Bergstrom (Minneapolis: University of Minnesota Press, 1991), 63–82, 64.

CHAPTER FIVE

1. "Odograph," "Is Peak Oil Pessimism a Generation of Men Coming to Realise How Useless They Are," Transition Culture, http://transitionculture .org/2006/12/04/is-peak-oil-pessimism-a-generation-of-men-coming-to -realise-how-useless-they-are.

2. "Ugo" Bardi (former president of the Association for the Study of Peak Oil), "Is Peak Oil Pessimism a Generation of Men Coming to Realise How Useless They Are," Transition Culture, http://transitionculture.org/2006/ 12/04/is-peak-oil-pessimism-a-generation-of-men-coming-to-realise-how -useless-they-are.

3. Ibid. (first three quotes); Kurt Cobb, "Is Peak Oil a Guy Thing?" Resource Insights, September 16, 2007, http://resourceinsights.blogspot.com/2007/ 09/is-peak-oil-guy-thing.html.

4. First applied to films, the Bechdel test identifies implicit sexism in media by whether a portrayal has "at least two women in it . . . who talk to each other . . . about something besides a man." Allison Bechdel, "The Rule," in *Dykes to Watch Out For* (Ithaca, NY: Firebrand, 1986).

5. Sharon Astyk, "Why Are the Mean Girls Picking on *World Made by Hand*?" The Chatelaine's Keys, May 4, 2008, http://sharonastyk.com/2008/ 05/04/kunstler-gets-defensive-meanderings-on-_world-made-by-hand _ (now defunct). On peak oil feminism, see Sharon Astyk, "Peak Oil Is Still a Women's Issue and Other Reflections on Sex, Gender and the Long Emergency," Casaubon's Blog, January 31, 2010, http://scienceblogs.com/ casaubonsbook/2010/01/31/peak-oil-is-still-a-womens-iss.

6. Kunstler, *World Made by Hand*, 101.

7. "Sublime Oblivion," "Book Review—*World Made by Hand*," http://www .sublimeoblivion.com/2009/04/10/notes-kunstler-fiction (page now deleted).

8. Janet McAdams, April 15, 2011, letter to Karla Armbuster in response to blog post on *World Made by Hand*, http://www.interversity.org/lists/asle/ archives/Apr2011/msg00074.html.

9. Astyk, "Why Are the Mean Girls Picking on *World Made by Hand*?"

10. "Re: What won't You Miss," Peak Oil News and Message Boards, July 22, 2008, post by "Cashmere," http://peakoil.com/forums/post732750.html.

11. "The Return of Patriarchy," Peak Oil News and Message Boards, February 14, 2010, http://peakoil.com/forums/the-return-of-patriarchy-t57793-20.html.
12. Gail Tverberg (posting as "Gail the Actuary"), "The Oil Drum: Campfire | Men's Response to Shifting Roles after Peak Oil," The Oil Drum, January 30, 2010, http://campfire.theoildrum.com/node/6171.
13. "The Return of Patriarchy," February 15, 2010.
14. Lindsay Hixson, Bradford B. Hepler, and Myoung Ouk Kim, "The White Population: 2010," Report no. C2010BR-05 (Washington, DC: US Census Bureau, September 2011).
15. See, e.g., Robert D. Bullard, *Dumping in Dixie: Race, Class and Environmental Quality* (Boulder, CO: Westview, 1990); Andrew Hurley, *Environmental Inequalities: Class, Race, and Industrial Pollution in Gary, Indiana, 1945–1980* (Chapel Hill: University of North Carolina Press, 1995); Dorceta E. Taylor, *Race, Class, Gender, and American Environmentalism* (Washington, DC: US Department of Agriculture, 2002); and David N. Pellow, *Garbage Wars: The Struggle for Environmental Justice in Chicago* (Cambridge, MA: MIT Press, 2004).
16. Paul B. Stretesky, Sheila Huss, Michael J. Lynch, Sammy Zahran, and Bob Childs, "The Founding of Environmental Justice Organizations across U.S. Counties during the 1990s and 2000s: Civil Rights and Environmental Cross-Movement Effects," *Social Problems* 58, no. 3 (2011): 330–60. In fact, sociologists have found that people of color are not only more vulnerable to but also more perceptive of existing environmental risks.
17. See Melissa L. Finucane, Paul Slovic, C. K. Mertz, James Flynn, and Theresa A. Satterfield, "Gender, Race, and Perceived Risk: The 'White Male' Effect," *Health, Risk and Society* 2 (2000): 159–72; and Brent K. Marshall, "Gender, Race, and Perceived Environmental Risk: The 'White Male' Effect in Cancer Alley, LA," *Sociological Spectrum* 24 (2000): 453–78.
18. Flynn, *Shut Down*, 234, 73, 127, 205. This aspect of the novel shows that the mind-set that leads to the kind of problematic connections that Malthusian environmentalists made between population control and the Global South in the 1960s and 1970s is still present, to some extent.
19. Respondents 9599590, 9597368.
20. James McCausland, "Scientists' Warnings Unheeded," *Brisbane Courier-Mail*, December 4, 2006, http://www.couriermail.com.au/business/scientists-warnings-unheeded/story-e6freqmx-1111112631991.
21. J. Emmett Winn, "Mad Max, Reaganism and the Road Warrior," *Kinema: A Journal for Film and Audiovisual Media* 23 (2009): 57–75.
22. The *Dirty Harry* series is *Dirty Harry* (Don Siegel, 1971), *Magnum Force* (Ted Post, 1973), *The Enforcer* (James Fargo, 1976), *Sudden Impact* (Clint Eastwood, 1983), and *The Dead Pool* (Buddy Van Horn, 1988). The *Death Wish* series is *Death Wish* (Michael Winner, 1974), *Death Wish II* (Michael Winner, 1982), *Death Wish 3* (Michael Winner, 1985), *Death Wish 4: The Crackdown* (J. Lee Thompson, 1987), and *Death Wish V: The Face of Death* (Allan A. Goldstein, 1994).

23. This paragraph, like most of this book, treats libertarianism and environmentalism as diametrically opposed perspectives, but, as Brian Allen Drake has shown, this was not always the case. For certain issues, a libertarian environmentalism has been not only possible but also coherent. See Brian Allen Drake, *Loving Nature, Fearing the State: Environmentalism and Antigovernment Politics Before Reagan* (Seattle: University of Washington Press, 2013). It is hard to imagine such an intersection on issues such as climate change, although anarchist thinkers have tried to picture one. See Peter Gelderloos, "An Anarchist Solution to Global Warming," 2010, http://theanarchistlibrary.org/library/Peter_Gelderloos__An_Anarchist _Solution_to_Global_Warming.html.

24. See Jenkins, *Decade of Nightmares*, 237–43; Wilentz, *The Age of Reagan*, 180–87; and Zaretsky, *No Direction Home*, 227–44.

25. Kunstler, *World Made by Hand*, 267.

26. Flynn, *Shut Down*, 239, 257.

27. This analysis assumes that masculinity is a social construct. While we should always be aware that there is never such a thing as a monolithic model of masculinity—different groups perform different masculinities on the basis of race, ethnicity, class, education, gender, sexual orientation, and other factors—I hold that mainstream popular culture plays a powerful role in forging hegemonic (or, as I put it in this chapter, *mainstream*) gender expectations, which are sometimes enforced by social, cultural, and economic actions. See R. W. Connell, "The Social Organization of Masculinity," in *The Masculinities Reader*, ed. Stephen M. Whitehead and Frank J. Barrett (Cambridge: Polity, 2001), 30–50.

28. Ronald E. Kutscher, "Historical Trends, 1950–92, and Current Uncertainties," *Monthly Labor Review*, November 1993, 3–10.

29. Sophie Quinton, "In Manufacturing, Blue-Collar Jobs Need White-Collar Training," *National Journal*, February 27, 2012, http://www.nationaljournal .com/whitehouse/in-manufacturing-blue-collar-jobs-need-white-collar -training-20120226.

30. Neil Irwin, "American Manufacturing Is Coming Back; Manufacturing Jobs Aren't," *Washington Post*, November 19, 2012, http://www .washingtonpost.com/blogs/wonkblog/wp/2012/11/19/american -manufacturing-is-coming-back-manufacturing-jobs-arent.

31. The film *Falling Down* (Joel Schumacher, 1993) would certainly be counted among this number had it been released a few years later.

32. Susan Faludi, *Stiffed: The Betrayal of the American Man* (New York: William Morrow, 1999), 6. Of course, wealthy white men had been complaining of the feminization of labor and the need for "authentic" physical experiences for over a century, since (at least) the shift from artisanal to industrial manufacturing. See T. J. Jackson Lears, *No Place of Grace: Antimodernism and the Transformation of American Culture, 1880–1920* (New York: Pantheon, 1981), esp. 59–96.

33. Theodore Roosevelt, lecture at Harvard University (1907), quoted in Kevin P. Murphy, *Political Manhood: Red Bloods, Mollycoddles, and the Politics of Progressive Era Reform* (New York: Columbia University Press, 2010), 1.

34. K. A. Cuordileone, *Manhood and American Political Culture in the Cold War* (Oxford: Routledge, 2005), 12.

35. Michael A. Messner, "The Limits of 'the Male Sex Role': An Analysis of the Men's Liberation and Men's Rights Movements' Discourse," *Gender and Society* 12, no. 3 (1998): 255–76, 266.

36. Andrew Romano and Tony Dokoupil, "Men's Lib," *Newsweek*, September 27, 2010, 43.

37. US Department of Labor, US Bureau of Labor Statistics, "Highlights of Women's Earnings in 2011," Report 1038, October 2012, http://www.bls.gov/cps/cpswom2011.pdf.

38. "The Oil Drum: Campfire," The Oil Drum, January 31, 2010, http://campfire.theoildrum.com/node/6171.

39. Faludi, *Stiffed*, 38.

40. Astyk, "Why Are the Mean Girls Picking On *World Made by Hand*?"

41. Toby Miller, "A Metrosexual Eye on *Queer Guy*," *GLQ: A Journal of Lesbian and Gay Studies* 11, no. 1 (2005): 112–17, 114.

42. *Queer Eye for the Straight Guy*, David Collins and David Metzler, Bravo, 2003–7.

43. See Alexandra Finkel, "Nail Polish for Men Is Latest Trend in Grooming," *New York Daily News*, November 3, 2011, http://www.nydailynews.com/life-style/fashion/nail-polish-men-latest-trend-grooming-article-1.971247#ixzz2UlJms272.

44. On retrosexuality among straight white men, see Romano and Dokoupil, "Men's Lib." On the "retrosexual code," see Tom Hawks, "The 50 Points of the Retrosexual/Neosexual Code, for Real Men Today," The Way One Vike Sees It, February 28, 2010, http://onevike.blogspot.com/2010/02/50-points-of-retrosexual-neosexual-code.html. On the broader phenomenon of recuperating previous gender roles, see Nishant Shahani, *Queer Retrosexualities: The Politics of Reparative Return* (Plymouth: Lehigh University Press, 2012).

45. See Ryan Malphurs, "The Media's Frontier Construction of President George W. Bush," *Journal of American Culture* 31, no. 2 (2008): 185–201.

46. Michael McAuliff, "Torture Like Jack Bauer's Would Be OK, Bubba Says," *New York Daily News*, October 1, 2007, 22.

47. Hamilton Carroll, *Affirmative Reaction: New Formations of White Masculinity* (Durham, NC: Duke University Press, 2011), 19.

48. Fire departments have remained one of the most segregated workplaces in the United States and have been the site of a number of lawsuits over reverse racism against white applicants. See Carol A. Chetkovich, *Real Heat: Gender and Race in the Urban Fire Service* (New Brunswick, NJ: Rutgers University Press, 1997).

49. See Richard Majors, "Cool Pose: Black Masculinity and Sports," in Whitehead and Barrett, eds., *The Masculinities Reader*, 209–18.

50. Lini S. Kadaba, "The New Retrosexual: Putting the Man Back in Manhood," *Philadelphia Inquirer*, April 23, 2010, http://www.capecodonline.com/apps/pbcs.dll/article?AID=/20100423/LIFE/100429929/-1/LIFE0707.

51. Dave Besley, *The Retrosexual Manual: How to Be a Real Man* (Chicago: Prion, 2008), front cover.

52. Aaron Traister, "'Retrosexuals': The Latest Lame Male Catchphrase," Salon, April 7, 2010, http://www.salon.com/2010/04/07/retrosexuals_silliness.

53. "Manly Skills," The Art of Manliness, n.d., http://www.artofmanliness.com/category/manly-skills; Brett McKay and Kate McKay, *The Art of Manliness: Classic Skills and Manners for the Modern Man* (Cincinnati: HOW, 2009).

54. Respondent 9591897.

55. Romano and Dokoupil, "Men's Lib," 43.

56. Stephanie Coontz, "Myth of Male Decline," *New York Times*, September 30, 2012, SR1.

57. Hanna Rosin, "Who Wears the Pants in This Economy?" *New York Times*, September 16, 2012, MM8.

58. Leo R. Chavez, *The Latino Threat: Constructing Immigrants, Citizens, and the Nation* (Stanford, CA: Stanford University Press, 2008), 2.

59. American National Election Studies, Time Series Study, 2008, http://www.thearda.com/Archive/Files/Analysis/NES2008/NES2008_Var1241_1.asp. See also Christopher P. Muste, "The Dynamics of Immigration Opinion in the United States, 1992–2012," *Public Opinion Quarterly* 77, no. 1 (2013): 398–416.

60. See, e.g., Heidi Shierholz, "Immigration and Wages—Methodological Advancements Confirm Modest Gains for Native Workers," Briefing Paper no. 255 (Washington, DC: Economic Policy Institute, February 4, 2010); and Elizabeth Dwoskin, "Why Americans Won't Do Dirty Jobs," *Bloomberg Businessweek*, November 9, 2011, http://www.businessweek.com/magazine/why-americans-wont-do-dirty-jobs-11092011.html#p2.

61. Jonathan Chait, "2012 or Never," *New York*, February 26, 2012, http://nymag.com/news/features/gop-primary-chait-2012-3.

62. See Kate Zernieke and Megan Thee-Brenan, "Discontent's Demography: Who Backs the Tea Party," *New York Times*, April 15, 2010, A1. For a thicker portrait of Tea Party activists, see Williamson, Skocpol, and Coggin, *The Tea Party and the Remaking of Republican Conservatism*, 3–82.

63. *Lost* (J.J. Abrams, 2004–10), *Iron Man* (Jon Favreau, 2008), and *Iron Man 2* (John Favreau, 2010). This analysis applies primarily to the first season of *Revolution*.

64. See, e.g., Torie Bosch, "The Biggest Problem with the Post-Apocalyptic Show *Revolution*: The Women's Perfect Hair," Slate, September 14, 2012, http://www.slate.com/blogs/future_tense/2012/09/14/revolution_the_post_apocalyptic_show_gets_the_hair_of_female_stars_tracy_spiridakos.html.

65. *Revolution*, NBC, season 1, episode 10, first aired on November 26, 2012.

66. Curtis, *Postapocalyptic Fiction and the Social Contract*, 5.
67. Kyle William Bishop, *Zombie Gothic: The Rise and Fall (and Rise) of the Walking Dead in Popular Culture* (Jefferson, NC: McFarland, 2010), 19.
68. Elizabeth Fox-Genovese, *Feminism without Illusions: A Critique of Individualism* (Chapel Hill: University of North Carolina Press, 1991), 122, 128.
69. See, e.g., George Lipsitz, *The Possessive Investment in Whiteness: How White People Profit from Identity Politics* (Philadelphia: Temple University Press, 2006).
70. Carol Pateman, *The Sexual Contract* (Stanford, CA: Stanford University Press, 1988); Charles W. Mills, *The Racial Contract* (Ithaca, NY: Cornell University Press, 1997).
71. Jack Turner, *Awakening to Race: Individualism and Social Consciousness in America* (Chicago: University of Chicago Press, 2012), 23.
72. Sally Robinson, *Marked Men: White Masculinity in Crisis* (New York: Columbia University Press, 2000), 3.
73. For other explorations of gender in the post-carbon future, see Margaret Atwood, *The Year of the Flood* (London: Bloomsbury, 2009); and Claire Colbrook, *Sex After Life*, vol. 2 of *Essays on Extinction* (Ann Arbor, MI: Open Humanities Press, 2014).

CONCLUSION

1. The average price of gasoline rose from $1.27 in January 2000 to a high of $4.11 in July 2008 and has mostly stayed in the $2.50–$4.00 range since then. Source: US Energy Information Administration, "Short-Term Energy Outlook," December 9, 2014, http://www.eia.gov/forecasts/steo/realprices.
2. US Energy Information Administration, "Primary Energy Overview," *Monthly Energy Review*, December 2014, http://www.eia.gov/totalenergy/data/monthly/pdf/sec1_3.pdf.
3. For hydrofacking, the environmental risk is in the production method itself, which many independent reports now blame for water contamination and an increase in earthquakes as well as methane leaks (methane is seventy times more potent than carbon dioxide as a greenhouse gas). See Michael Behar, "Fracking's Latest Scandal? Earthquake Swarms," *Mother Jones*, March/April 2013, http://www.motherjones.com/environment/2013/03/does-fracking-cause-earthquakes-wastewater-dewatering; David C. Holzman, "Methane Found in Well Water Near Fracking Sites," *Environmental Health Perspectives* 119, no. 7 (2011): A289; Seongeun Jeong, Ying-Kuang Hsu, Arlyn E. Andrews, Laura Bianco, Patrick Vaca, James M. Wilczak, and Marc L. Fischer, "A Multitower Measurement Network Estimate of California's Methane Emissions," *Journal of Geophysical Research: Atmospheres* 118, no. 19 (2013): 11–339; and Robert W. Howarth, Anthony Ingraffea, and Terry Engelder, "Natural Gas: Should Fracking Stop?" *Nature* 477, no. 7364 (2011): 271–75. For tar sands, the environmental issue is also

the scale of the operation as well as the potential emissions: tar sands' low EROI means that the emissions are approximately 14 percent higher than they are for conventional crude oil. See Richard K. Lattanzio, "Canadian Oil Sands: Life-Cycle Assessments of Greenhouse Gas Emissions" (Washington, DC: Congressional Research Service, March 10, 2014), http://fas .org/sgp/crs/misc/R42537.pdf; and David Biello, "How Much Will Tar Sands Oil Add to Global Warming?" *Scientific American*, January 23, 2013, http://www.scientificamerican.com/article/tar-sands-and-keystone-xl -pipeline-impact-on-global-warming.

4. "The Last Post," The Oil Drum, September 22, 2013, http://www .theoildrum.com/node/10249?utm_source=feedburner&utm_medium= feed&utm_campaign=Feed%3A+theoildrum+%28The+Oil+Drum%29& utm_content=FeedBurner.

5. Liam Denning, "Has Peak Oil Peaked?" *Wall Street Journal* blog, June 26, 2012, http://blogs.wsj.com/overheard/2012/06/26/has-peak-oil-peaked.

6. David Frum, "'Peak Oil' Doomsayers Proved Wrong," CNN Opinion, March 4, 2013, http://www.cnn.com/2013/03/04/opinion/frum-peak-oil; David Blackmon, "As Fracking Rises, Peak Oil Theory Slowly Dies," *Forbes*, July 16, 2013, http://www.forbes.com/sites/davidblackmon/2013/07/16/as -fracking-rises-peak-oil-theory-slowly-dies.

7. Methane hydrates (also known as methane clathrates) are compounds in which methane is trapped in ice. They can be found in huge amounts in subsea deposits on the continental shelf, and the methane might someday be extracted and used for fuel. See Charles C. Mann, "What If We Never Run Out of Oil?" *Atlantic*, April 24, 2013, 48–63.

8. Respondents 9682027, 9682027, 9686312, 9700344, 9680437, 9679616.

9. Respondent 9594065; Brian Henderberg; respondent 9601806; respondent 9687367.

10. Respondents 9711257, 9585147, 9687367.

11. Respondents 9574313, 9577042, 9596820, 9582295, 9679616.

12. Respondents 9700344, 9601806.

13. See, e.g., Roy Scranton, "Learning How to Die in the Anthropocene," *New York Times*, November 10, 2013, http://opinionator.blogs.nytimes .com/2013/11/10/learning-how-to-die-in-the-anthropocene; and many of the essays in Tom Cohen, ed., *Telemorphosis: Theory in the Era of Climate Change*, 1 vol. to date (n.p.: Open Humanities Press, 2012).

14. Respondents 9577182, 9695655.

15. Buell, *From Apocalypse to Way of Life*.

16. David Biello, "400 PPM: Can Artificial Trees Help Pull CO2 from the Air?" *Scientific American*, May 9, 2013, http://www.scientificamerican.com/ article.cfm?id=prospects-for-direct-air-capture-of-carbon-dioxide.

17. John Vidal, "Global Warming Causes 300,000 Deaths a Year, Says Kofi Annan Thinktank," *Guardian*, May 29, 2009, http://www.theguardian .com/environment/2009/may/29/1. This number is calculated as the result

of floods, droughts, fires, new diseases, and small-scale warfare that can be attributed to anthropogenic climate change. By 2030, this number is expected to be as high as 500,000 people per year.

18. Eleanor J. Burke, Simon J. Brown, and Nikolaos Christidis, "Modeling the Recent Evolution of Global Drought and Projections for the Twenty-First Century with the Hadley Centre Climate Model," *Journal of Hydrometeorology* 7, no. 5 (2006): 1113–25.

19. Jennifer Leaning and Debarati Guha-Sapir, "Natural Disasters, Armed Conflict, and Public Health," *New England Journal of Medicine* 369 (2013): 1836–42.

20. Tom Levitt, "Overfished and Under-Protected: Oceans on the Brink of Catastrophic Collapse," CNN, March 27, 2013, http://www.cnn.com/2013/03/22/world/oceans-overfishing-climate-change. See also Erik Vance, "Letter from the Sea of Cortez: Emptying the World's Aquarium; The Dismal Future of the Global Fishery," *Harper's*, August 2013, 53–62.

21. Sigmar Gabriel, "Biodiversity 'Fundamental' to Economics," BBC News, March 9, 2007, http://news.bbc.co.uk/2/hi/science/nature/6432217.stm.

22. Garry Peterson, "Visualizing the Great Acceleration—Part II," Resilience Science, December 4, 2008, http://rs.resalliance.org/2008/12/04/visualizing-the-great-acceleration-part-ii. Data from William L. Steffen, Angelina Sanderson, Peter Tyson, Jill Jäger, Pamela A. Matson, Berrien Moore III, and Frank Oldfield, *Global Change and the Earth System: A Planet under Pressure* (Berlin: Springer, 2005).

23. Carl Boggs, *Ecology and Revolution: Global Crisis and the Political Challenge* (New York: Palgrave Macmillan, 2012), 32–33.

24. World Commission on Environment and Development, *Our Common Future* (Oxford: Oxford University Press, 1987), 47. This is often known as the Brundtland Commission after its chairman, Gro Harlem Brundtland, then the prime minister of Norway. See Daniel J. Philippon, "Sustainability and the Humanities: An Extensive Pleasure," *American Literary History* 24, no. 1 (2012): 163–79.

25. Indeed, some claim that, in its current meaning, *sustainable development* was first used in *Limits to Growth*. See Ulrich Grober, "Deep Roots: A Conceptual History of 'Sustainable Development'" (master's thesis, Wissenschaftszentrum Berlin für Sozialforschung, 2007), 8–9.

26. Mathis Wackernagel and William Rees, *Our Ecological Footprint: Reducing Human Impact on the Earth* (Gabriola Island, BC: New Society, 1996), 9.

27. Global Footprint Network, "Data Sources," n.d., http://www.footprintnetwork.org/en/index.php/GFN/page/data_sources. An ecological footprint is a calculation that estimates the area of Earth's productive land and water required to supply the resources that an individual or group demands as well as to absorb the wastes that the individual or group produces.

28. The meaning of *sustainable* and that of the concept of sustainability

have been legitimately challenged by many environmental scholars. See, e.g., Donald Worster, *The Wealth of Nature: Environmental History and the Ecological Imagination* (New York: Oxford University Press, 1993), 142–55. Whatever more precise and less malleable paradigm emerges from this ongoing debate, it is clear that our current rates of consumption, energy use, and waste production are not sustainable for any long period of time. See, e.g., Speth, *The Bridge at the End of the World*, 17–45.

29. Buell, *From Apocalypse to Way of Life*, 74. For more information on each of these issues, see ibid., 69–142.

30. The possible exception here is the hole in the ozone over the Earth's polar regions, which was largely caused by the dissociation of chemicals such as CFCs. The Montreal Protocol on Substances that Deplete the Ozone Layer, signed internationally in 1989, successfully restricted the production of these chemicals and remains one of the great victories for environmentalism internationally. However, it will be decades before the hole in the stratosphere recovers, and scientists are just beginning to understand the interaction between ozone depletion and climate change. See Son Seok-Woo, Neil F. Tandon, Lorenzo M. Polvani, and Darryn W. Waugh, "Ozone Hole and Southern Hemisphere Climate Change," *Geophysical Research Letters* 36, no. 15 (2009): L15705.

31. Sarah S. Amsler, "Bringing Hope 'to Crisis': Crisis Thinking, Ethical Action and Social Change," in Skrimshire, ed., *Future Ethics*, 129–52, 129.

32. Michael Pollan, "Why Bother?" *New York Times*, April 20, 2008, MM19.

33. Pew Research Center for the People and the Press, "Deficit Reduction Rises on Public's Agenda for Obama's Second Term," January 24, 2013, http://www.people-press.org/2013/01/24/deficit-reduction-rises-on-publics-agenda-for-obamas-second-term.

34. Buell, *From Apocalypse to Way of Life*, 39.

35. *Tough oil* is a term for petroleum with a higher EROI than was found in the past—it is "tough," i.e., expensive and risky to exploit, because it is (for the most part) subsea, deep underground, and/or suspended in complex geologic formations, such as shale oil and tar sands.

36. Respondent 9593771.

37. For example, a 2009 report by the McKinsey Global Institute suggested that available conservation measures could cut residential energy use (which makes up one-quarter of global energy use) in half. McKinsey and Co., *Curbing Global Energy-Demand Growth: The Energy Productivity Opportunity*, by Florian Bressand, Diana Farrell, Pedro Haas, Fabrice Morin, Scott Nyquist, Jaana Remes, Sebastian Roemer, Matt Rogers, Jaeson Rosenfeld, and Jonathan Woetzel (n.p.: McKinsey Global Institute, 2007), http://go.nature.com/226nPx.

38. Norgaard, *Living in Denial*, 193.

39. See, e.g., Vladas Griskevicius, Joshua M. Tybur, and Bram Van den Bergh, "Going Green to Be Seen: Status, Reputation, and Conspicuous Conserva-

tion," *Journal of Personality and Social Psychology* 98, no. 3 (2010): 392–404, 392; Niamh Murtagh, Birgitta Gatersleben, and David Uzzell, "Multiple Identities and Travel Mode Choice for Regular Journeys," *Transportation Research Part F: Traffic Psychology and Behaviour* 15, no. 55 (2012): 514–24; and Hal E. Hershfield, H. Min Bang, and Elke U. Weber, "National Differences in Environmental Concern and Performance Are Predicted by Country Age," *Psychological Science* 25, no. 1 (2014): 152–60.

40. Respondents 9601806, 9686312.

41. Nina Eliasoph, *Avoiding Politics: How Americans Produce Apathy in Everyday Life* (Cambridge: Cambridge University Press, 1998).

42. Respondent 9575985 (emphasis added).

43. See, e.g., Naomi Oreskes and Erik Conway, *Merchants of Doubt: How a Handful of Scientists Obscured the Truth on Issues from Tobacco Smoke to Global Warming* (New York: Bloomsbury, 2010); and James Hoggan, *Climate Cover-Up: The Crusade to Deny Global Warming* (Vancouver, BC: Greystone, 2009).

44. Susan Opotow and Leah Weiss, "New Ways of Thinking about Environmentalism: Denial and the Process of Moral Exclusion in Environmental Conflict," *Journal of Social Issues* 56, no. 3 (2000), 443–57, 477.

45. Stanley Cohen, *States of Denial: Knowing about Atrocities and Suffering* (Cambridge: Polity, 2001), 8.

46. Norgaard, *Living in Denial*, 5, 92, 9.

47. As Ehrenreich notes, the "cult of cheerfulness" is powerful exactly not because it is in no way an official mandate but because it is seemingly reproduced and demanded by each individual citizen. Ehrenreich, *Bright-Sided*, 51. One could very well describe the pressures to conform to the expectations of the dominant social paradigm through the lens of Michel Foucault's concept of governmentality.

48. Eddie Yuen, "The Politics of Failure Have Failed: The Environmental Movement and Catastrophism," in *Catastrophism: The Apocalyptic Politics of Collapse and Rebirth*, by Sasha Lilley, David McNally, Eddie Yuen, and James Davis (Oakland, CA: PM, 2012), 15–43, 21.

49. Rodgers, *The Age of Fracture*, 41.

50. See Adam Serwer, Jaeah Lee, and Zaineb Mohammed, "Now That's What I Call Gerrymandering!" *Mother Jones*, November 14, 2012, http://www.motherjones.com/politics/2012/11/republicans-gerrymandering-house-representatives-election-chart.

51. North, "Eco-Localisation as a Progressive Response," 586.

52. See, e.g., Daniel Yergin, "There Will Be Oil," *Wall Street Journal*, September 17, 2011, C1.

53. On the failure of emissions trading in the implementation of the 1977 Clean Air Act Amendments, see David M. Driesen, "Neoliberal Instrument Choice," in *Economic Thought and US Climate Change Policy*, ed. David M. Driesen (Cambridge: MIT Press, 2010), 129–50. On the failure of contem-

porary European emissions markets, see Nathaniel Gronewold, "Europe's Carbon Emissions Trading—Growing Pains or Wholesale Theft?" *New York Times*, January 31, 2011, http://www.nytimes.com/cwire/2011/01/31/ 31climatewire-europes-carbon-emissions-trading-growing-pai-74999.html ?pagewanted=all; and Elisabeth Rosenthal and Andrew W. Lehren, "Profits on Carbon Credits Drive Output of a Harmful Gas," *New York Times*, August 8, 2012.

54. Nicholas Stern, *Review on the Economics of Climate Change* (London: HM Treasury, 2006), 11. On the influence of libertarian-neoliberal thought on global environmental policy, see Driesen, "Neoliberal Instrument Choice"; and Jane Andrew, Mary A. Kaidonis, and Brian Andrew, "Carbon Tax: Challenging Neoliberal Solutions to Climate Change," *Critical Perspectives on Accounting* 21, no. 7 (2010): 611–18.

55. Respondent 9686274.

Bibliography

Unless otherwise noted, all URLs were current and active when this book went to press.

Abbate, Janet. *Inventing the Internet*. Cambridge, MA: MIT Press, 1999.

Abdel-Khalek, Ahmed, and David Lester. "Optimism and Pessimism in Kuwaiti and American College Students." *International Journal of Social Psychiatry* 52, no. 2 (2006): 110–26.

Akins, James. "The Oil Crisis: This Time the Wolf Is Here." *Foreign Affairs* 51, no. 3 (1973): 462–90.

Alexander, Michelle. *The New Jim Crow: Mass Incarceration in the Age of Colorblindness*. New York: New Press, 2010.

Alfano, Sean. "Poll: Gas Prices Affecting Habits." CBS News, September 1, 2005. http://www.cbsnews.com/news/poll-gas -prices-affecting-habits.

American Petroleum Institute. *Petroleum Facts and Figures*. New York, 1971.

Amsler, Sarah S. "Bringing Hope 'to Crisis': Crisis Thinking, Ethical Action and Social Change." In *Future Ethics: Climate Change and Apocalyptic Imagination*, ed. Stefan Skrimshire, 129–52. London: Continuum, 2010.

Anderson, Benedict. *Imagined Communities: Reflections on the Spread of Nationalism*. London: Verso, 1983.

Anderson, Christopher. "The Accidental Superhighway." *Economist*, July 1, 1995, 5–20.

Andrew, Jane, Mary A. Kaidonis, and Brian Andrew. "Carbon Tax: Challenging Neoliberal Solutions to Climate Change." *Critical Perspectives on Accounting* 21, no. 7 (2010): 611–18.

Appelbaum, Binyamin, and Robert Gebeloff. "Even Critics of Safety Net Increasingly Depend on It." *New York Times*, February 12, 2012, A1.

Astyk, Sharon. "Why Are the Mean Girls Picking on *World Made by Hand*?" The Chatelaine's Keys, May 4, 2008. http://sharonastyk.com/2008/05/ 04/kunstler-gets-defensive-meanderings-on-_world-made-by-hand_ (now defunct).

———. "Peak Oil Is Still a Women's Issue and Other Reflections on Sex, Gender and the Long Emergency." Casaubon's Book, January 31, 2010. http:// scienceblogs.com/casaubonsbook/2010/01/31/peak-oil-is-still-a-womens-iss/.

Atkinson, Adrian. "Progress in Acknowledging and Confronting Climate Change and 'Peak Oil.'" *City: Analysis of Urban Trends, Culture, Theory, Policy, Action* 14, no. 3 (2010): 314–22.

Atwood, Margaret. *The Year of the Flood*. London: Bloomsbury, 2009.

Ausubel, Herman. *Historians and Their Craft: A Study of the Presidential Addresses of the American Historical Association, 1884–1945*. New York: Russell & Russell, 1965.

Auxier, Randall E. "Foucault, Dewey, and the History of the Present." *Journal of Speculative Philosophy* 16, no. 2 (2002): 75–102.

Bailey, Ian, Rob Hopkins, and Geoff Wilson. "Some Things Old, Some Things New: The Spatial Representations and Politics of Change of the Peak Oil Relocalisation Movement." *Geoforum* 41, no. 4 (2010): 595–605.

Baker, Carolyn. *Sacred Demise: Walking the Path of Industrial Civilization's Collapse*. Bloomington, IN: iUniverse, 2009.

Baker, Wayne. *America's Crisis of Values: Reality and Perception*. Princeton, NJ: Princeton University Press, 2005.

Ball, Jeffrey. "As Prices Soar, Doomsayers Provoke Debate on Oil's Future." *Wall Street Journal*, September 21, 2004, A1.

Ballantyne, A. P., C. B. Alden, J. B. Miller, P. P. Tans, and J. W. C. White. "Increase in Observed Net Carbon Dioxide Uptake by Land and Oceans during the Past 50 Years." *Nature* 488, no. 7409 (2012): 70–72.

Balmford, Andrew, Linda Birkin, Andrea Manica, Lesley Airey, Amy Oliver, and Judith Schleicher. "Hollywood, Climate Change, and the Public." *Science* 305, no. 5691 (2004): 1713.

Bandura, Albert. "Exercise of Human Agency through Collective Efficacy." *Current Directions in Psychological Science* 9, no. 3 (2000): 75–78.

Bardi, Ugo. "Peak Oil: the Four Stages of a New Idea." *Energy* 34, no. 3 (2009): 323–26.

Bardini, Thierry. *Bootstrapping: Douglas Engelbart, Coevolution, and the Origins of Personal Computing*. Stanford, CA: Stanford University Press, 2000.

Barkun, Michael. "Millennium Culture: The Year 2000 as a Religious Event." In *Millennial Visions: Essays on Twentieth-century Millenarianism*, ed. Martha F. Lee, 41–54. Westport, CT: Praeger, 2000.

Barney, Darin. *The Network Society*. Malden, MA: Polity, 2004.

Barrett, Brendan. "Reality of Peak Oil Enters Our Fiction." Our World 2.0, January 21, 2011. http://ourworld.unu.edu/en/reality-of-peak-oil-enters-our -fiction.

Bartolini, Stefano, Ennio Bilancini, and Maurizio Pugno. "Did the Decline in Social Capital Decrease American Happiness? A Relational Explanation of the Happiness Paradox." *Social Indicators Research* 110, no. 3 (2013): 1033–59.

Beaton, Kendall. *Enterprise in Oil: A History of Shell in the United States.* New York: Appleton-Century-Crofts, 1957.

Bechdel, Allison. "The Rule." In *Dykes to Watch Out For.* Ithaca, NY: Firebrand, 1986.

Becken, Susan. "Developing Indicators for Managing Tourism in the Face of Peak Oil." *Tourism Management* 29, no. 4 (2008): 695–705.

Behar, Michael. "Fracking's Latest Scandal? Earthquake Swarms." *Mother Jones,* March/April 2013, http://www.motherjones.com/environment/2013/03/does-fracking-cause-earthquakes-wastewater-dewatering.

Bellah, Robert N., Richard Madsen, William M. Sullivan, Ann Swidler, and Steven M. Tipton. *Habits of the Heart: Individualism and Commitment in American Life.* Berkeley and Los Angeles: University of California Press, 1985.

Bendle, Mervyn F. "The Apocalyptic Imagination and Popular Culture." *Journal of Religion and Popular Culture* 11 (Fall 2005): 1–42.

Bendzela, Mike. "The End of the End: How the Peak Oil Movement Failed." OpEdNews, January 2, 2011. http://www.opednews.com/articles/The-End-of-The-End—How-t-by-Mike-Bendzela-110102-802.html.

Bennett, W. Lance, Regina G. Lawrence, and Steven Livingston. *When the Press Fails: Political Power and the News Media from Iraq to Katrina.* Chicago: University of Chicago Press, 2007.

Berlatsky, Noah, ed. *The Exxon Valdez Oil Spill.* Farmington Hill, MI: Greenhaven, 2011.

Besley, Dave. *The Retrosexual Manual: How to Be a Real Man.* Chicago: Prion, 2008.

Biello, David. "400 PPM: Can Artificial Trees Help Pull CO2 from the Air?" *Scientific American,* May 9, 2013. http://www.scientificamerican.com/article.cfm?id=prospects-for-direct-air-capture-of-carbon-dioxide.

———. "How Much Will Tar Sands Oil Add to Global Warming?" *Scientific American,* January 23, 2013. http://www.scientificamerican.com/article/tar-sands-and-keystone-xl-pipeline-impact-on-global-warming.

Bijker, Wiebe E., Thomas P. Hughes, and Trevor J. Pinch, eds. *The Social Construction of Technological Systems: New Directions in the Sociology and History of Technology.* Cambridge, MA: MIT Press, 1987.

Bishop, Kyle William. *Zombie Gothic: The Rise and Fall (and Rise) of the Walking Dead in Popular Culture.* Jefferson, NC: McFarland, 2010.

Black, Brian C. *Crude Reality: Petroleum in World History.* New York: Rowman & Littlefield, 2012.

———. "Oil for Living: Petroleum and American Conspicuous Consumption." *Journal of American History* 99, no. 1 (2012): 40–50.

Black, Edwin. *Internal Combustion: How Corporations and Governments Addicted the World to Oil and Derailed the Alternatives.* New York: St. Martin's, 2006.

Blackmon, David. "As Fracking Rises, Peak Oil Theory Slowly Dies." *Forbes,* July 16, 2013. http://www.forbes.com/sites/davidblackmon/2013/07/16/as -fracking-rises-peak-oil-theory-slowly-dies.

Boggs, Carl. *The End of Politics: Corporate Power and the Decline of the Public Sphere.* New York: Guilford, 2000.

———. "Warrior Nightmares: American Reactionary Populism at the Millennium." *Socialist Register* 36 (2000): 243–56.

———. *Ecology and Revolution: Global Crisis and the Political Challenge.* New York: Palgrave Macmillan, 2012.

Booth, Jim. *Boil Over: The Day the Oil Ran Out.* Charleston, SC: CreateSpace, 2010.

Borsook, Paulina. *Cyberselfish: A Critical Romp through the Terrible Libertarian Culture of High Tech.* New York: PublicAffairs, 2000.

Bosch, Torie. "The Biggest Problem with the Post-Apocalyptic Show *Revolution*: The Women's Perfect Hair." Slate, September 14, 2012. http://www.slate .com/blogs/future_tense/2012/09/14/revolution_the_post_apocalyptic _show_gets_the_hair_of_female_stars_tracy_spiridakos.html.

Boulianne, Shelley. "Does Internet Use Affect Engagement? A Meta-Analysis of Research." *Political Communication* 26, no. 2 (2009): 193–211.

Boyer, Paul. *By the Bomb's Early Light.* New York: Pantheon, 1985.

Braedley, Susan, and Meg Luxton, eds. *Neoliberalism and Everyday Life.* Montreal: McGill-Queens University Press, 2010.

Brinkley, Alan. "The New Deal and the Idea of the State." In *The Rise and Fall of the New Deal Order, 1930–1980,* ed. Steve Fraser and Gary Gerstle, 85–121. Princeton, NJ: Princeton University Press, 1989.

Brodwin, David. "Americans Leading a 'Do It Yourself Economy' as Washington Stalls." *U.S. News and World Report,* August 2, 2012. http://www.usnews .com/opinion/blogs/economic-intelligence/2012/08/02/americans-leading -a-do-it-yourself-economy-as-washington-stalls.

Brooks, Max. *The Zombie Survival Guide: Complete Protection from the Living Dead.* New York: Broadway, 2003.

Brown, Paul. "The Deathwatch for Cheap Oil." *New York Times,* October 14, 2006, C5.

Buell, Frederick. *From Apocalypse to Way of Life: Environmental Crisis in the American Century.* New York: Routledge, 2004.

Bullard, Robert D. *Dumping in Dixie: Race, Class and Environmental Quality.* Boulder, CO: Westview, 1990.

Burke, Eleanor J., Simon J. Brown, and Nikolaos Christidis. "Modeling the Recent Evolution of Global Drought and Projections for the Twenty-First Century with the Hadley Centre Climate Model." *Journal of Hydrometeorology* 7, no. 5 (2006): 1113–25.

Bury, J. B. *The Idea of Progress: An Inquiry into its Origins and Growth.* Cambridge: Cambridge University Press, 1920.

Butler, Octavia. *Parable of the Sower.* New York: Four Walls Eight Windows, 1993.

Buttel, Frederick H., Ann P. Hawkins, and Alison G. Power. "From Limits to Growth to Global Change: Constraints and Contradictions in the Evolution of Environmental Science and Ideology." *Global Environmental Change* 1, no. 1 (1990): 57–66.

Campbell, Colin J. "The End of Cheap Oil." *Scientific American* 278, no. 3 (1998): 60–65.

———. *The Coming Oil Crisis.* Brentwood: Multi-Science, 2004.

Cape, John M., and Laura Buckner. *Oil Dusk: A Peak Oil Story.* Humble, TX: Singing Bowl, 2009.

Caputi, Jane. "*Jaws* as Patriarchal Myth." *Journal of Popular Film* 6, no. 4 (1978): 305–26.

Carlson, Scott. "A Social Critic Warns of Upheavals to Come." *Chronicle of Higher Education*, October 20, 2006. http://chronicle.com/article/A-Social -Critic-Warns-of/15941.

Carr, Nicholas. *The Shallows: What the Internet Is Doing to Our Brains.* New York: Norton, 2010.

Carroll, Hamilton. *Affirmative Reaction: New Formations of White Masculinity.* Durham, NC: Duke University Press, 2011.

Carroll, Peter N. *It Seemed Like Nothing Happened: America in the 1970s.* New Brunswick, NJ: Rutgers University Press, 2000.

Castells, Manuel. *The Rise of the Network Society.* Cambridge, MA: Blackwell, 1996.

Cavanor, Natalie. "Running on Empty; Two Video Producers Paint a Bleak Picture of Oil Wars and the End of the American Suburbs." *New York Times*, March 13, 2005, LI1.

Cerulo, Karen A. *Never Saw It Coming: Cultural Challenges to Envisioning the Worst.* Chicago: University of Chicago Press, 2006.

Chait, Jonathan. "2012 or Never." *New York*, February 26, 2012. http://nymag .com/news/features/gop-primary-chait-2012-3.

Chan, Sue. "Massive Anti-War Outpouring." CBS News, February 16, 2003. http://www.cbsnews.com/news/massive-anti-war-outpouring.

Chandler, Graham. "The Cult of Peak Oil." *Alberta Oil: The Business of Energy*, October 1, 2007. http://www.albertaoilmagazine.com/2007/10/the-cult-of -peak-oil.

Chang, Edward C., Kiyoshi Asakawa, and Lawrence J. Sanna. "Cultural Variations in Optimistic and Pessimistic Bias: Do Easterners Really Expect the Worst and Westerners Really Expect the Best When Predicting Future Life Events?" *Journal of Personality and Social Psychology* 81, no. 3 (2001): 476–91.

Chavez, Leo R. *The Latino Threat: Constructing Immigrants, Citizens, and the Nation.* Stanford, CA: Stanford University Press, 2008.

Chesters, Graeme, and Ian Walsh. *Complexity and Social Movements: Multitudes at the Edge of Chaos.* New York: Routledge, 2006.

Chetkovich, Carol A. *Real Heat: Gender and Race in the Urban Fire Service.* New Brunswick, NJ: Rutgers University Press, 1997.

Clark, David D. "A Cloudy Crystal Ball—Visions of the Future." Presentation given at the Internet Engineering Task Force meeting, July 16, 1992. http://www.ietf.org/old/2009/proceedings/prior29/IETF24.pdf.

Clark, Duncan. "UK Will Face Peak Oil Crisis within Five Years, Report Warns." *Guardian*, October 29, 2008, http://www.theguardian.com/environment/2008/oct/29/fossil-fuels-oil.

Cleveland, C. J. "Net Energy from Oil and Gas Extraction in the United States, 1954–1997." *Energy* 30, no. 5 (2005): 769–82.

Cleveland, C. J., R. Costanza, C. A. S. Hall, and R. Kaufmann. "Energy and the U.S. Economy: A Biophysical Perspective." *Science* 225 (1984): 890–97.

Cleveland, Cutler J., and Peter A. O'Connor. "Energy Return on Investment of Oil Shale." *Sustainability* 3 (2011): 2307–33.

Clinton, William J., and Albert Gore Jr. "A Framework for Global Electronic Commerce." July 1, 1997. http://clinton4.nara.gov/WH/New/Commerce/read.html.

Cobb, Kurt. "Is Peak Oil a Guy Thing?" Resource Insights, September 16, 2007. http://resourceinsights.blogspot.com/2007/09/is-peak-oil-guy-thing.html.

———. "What Should Members of the Peak Oil Movement Call Themselves?" Energy Bulletin, December 2, 2007. http://www.energybulletin.net/node/38052.

———. *Prelude*. Pittsburgh, PA: Public Interest Communications, 2010.

Coffin, Raymond J., and Mark W. Lipsey. "Moving Back to the Land: An Ecologically Responsible Lifestyle Change." *Environment and Behavior* 13, no. 1 (1981): 42–63.

Cohen, Lizabeth. *A Consumer's Republic: The Politics of Mass Consumption in Postwar America*. New York: Vintage, 2003.

Cohen, Stanley. *States of Denial: Knowing about Atrocities and Suffering*. Cambridge: Polity, 2001.

Cohen, Tom, ed. *Telemorphosis: Theory in the Era of Climate Change*. 1 vol. to date. n.p.: Open Humanities Press, 2012.

Colborn, Theo, Dianne Dumanoski, and John Peter Meyes. *Our Stolen Future: Are We Threatening Our Fertility, Intelligence, and Survival?* New York: Plume, 1997.

Colbrook, Claire. *Sex After Life*. Vol. 2 of *Essays on Extinction*. Ann Arbor, MI: Open Humanities Press, 2014.

Collins, Robert M. *More: The Politics of Economic Growth in the Postwar World*. New York: Oxford University Press, 2000.

"Combien de pétrole dans mon ordinateur?" Actualité, n.d. http://www.linternaute.com/actualite/savoir/07/petrole-yaourt/7.shtml.

Connell, R. W. "The Social Organization of Masculinity." In *The Masculinities Reader*, ed. Stephen M. Whitehead and Frank J. Barrett, 30–50. Cambridge: Polity, 2001.

Cook, David A. *Lost Illusions: American Cinema in the Shadow of Watergate and Vietnam, 1970–1979*. New York: Scribner, 1999.

Cooke, Ronald R. *Oil, Jihad and Destiny.* New York: Opportunity Analysis, 2004.

Coontz, Stephanie. *The Way We Never Were: American Families and the Nostalgia Trap.* New York: Basic, 2000.

———. "Myth of Male Decline." *New York Times,* September 30, 2012, SR1.

Corcoran, Terence. "The 'Peak Oil' Cult." *National Post* (Don Mills, ON), October 12, 2004, FP15.

Coupland, Douglas. *Player One: What Is to Become of Us.* Toronto: House of Anansi, 2010.

Couri, Rick. "Poll Places Trust in Congress below Telemarketers." Newstalk Radio KRMG, December 13, 2011. http://www.krmg.com/news/news/local/poll-places-trust-congress-below-telemarketers/nFz8m.

Cowie, Jefferson. *Stayin' Alive: The 1970s and the Last Days of the Working Class.* New York: New Press, 2010.

Crespino, Joseph. *In Search of Another Country: Mississippi and the Conservative Counterrevolution.* Princeton, NJ: Princeton University Press, 2007.

Critchlow, Donald T. *Phyllis Schlafly and Grassroots Conservatism: A Woman's Crusade.* Princeton, NJ: Princeton University Press, 2005.

Cuordileone, K. A. *Manhood and American Political Culture in the Cold War.* Oxford: Routledge, 2005.

Curtis, Claire P. *Postapocalyptic Fiction and the Social Contract: "We'll Not Go Home Again."* Lanham, MD: Rowman & Littlefield, 2012.

Curtis, Fred. "Peak Globalization: Climate Change, Oil Depletion and Global Trade." *Ecological Economics* 69, no. 2 (2009): 427–34.

Darley, Julian. *High Noon for Natural Gas.* White River Junction, VT: Chelsea Green, 2004.

DeBord, Matthew. Review of *$20 Per Gallon* by Christopher Steiner. *Los Angeles Times,* May 26, 2009. http://articles.latimes.com/2009/jul/26/entertainment/ca-christopher-steiner26

Deffeyes, Kenneth S. *Beyond Oil: The View from Hubbert's Peak.* New York: Hill & Wang, 2005.

de Landa, Manuel. *A New Philosophy of Society: Assemblage Theory and Social Complexity.* New York: Continuum, 2006.

Denby, David. "Lava Story." *New York,* May 12, 1997, 58.

Denning, Liam. "Has Peak Oil Peaked?" *Wall Street Journal* blog, June 26, 2012. http://blogs.wsj.com/overheard/2012/06/26/has-peak-oil-peaked.

Diamond, Jared. *Guns, Germs and Steel: The Fates of Human Societies.* New York: Norton, 1999.

———. *Collapse: How Societies Choose to Fail or Succeed.* New York: Viking, 2005.

Dillon, Merton L. "The Failure of the American Abolitionists." *Journal of Southern History* 25, no. 2 (1959): 159–77.

Doherty, Brian. *Radicals for Capitalism: A Freewheeling History of the Modern American Libertarian Movement.* New York: PublicAffairs, 2007.

Douglas, Richard McNeill. "The Ultimate Paradigm Shift: Environmentalism as Antithesis to the Modern Paradigm of Progress." In *Future Ethics: Climate*

Change and Apocalyptic Imagination, ed. Stefan Skrimshire, 197–218. New York: Continuum, 2010.

Drake, Brian Allen. *Loving Nature, Fearing the State: Environmentalism and Anti-government Politics Before Reagan*. Seattle: University of Washington Press, 2013.

Driesen, David M. "Neoliberal Instrument Choice." In *Economic Thought and US Climate Change Policy*, ed. David M. Driesen, 129–50. Cambridge, MA: MIT Press, 2010.

Drury, John, Christopher Cocking, Joseph Beale, Charlotte Hanson, and Faye Rapley. "The Phenomenology of Empowerment in Collective Action." *British Journal of Social Psychology* 44, no. 3 (2005): 309–28.

Drury, John, and Steve Reicher. "Collective Psychological Empowerment as a Model of Social Change: Researching Crowds and Power." *Journal of Social Issues* 65, no. 4 (2009): 707–25.

Dryzek, John S. *The Politics of the Earth: Environmental Discourses*. Oxford: Oxford University Press, 2012.

Duffey, Michael R. "Prognosticating Oil Supplies: Lower Reserves and Higher Prices Ahead?" *Washington Times*, November 3, 2004, A23.

Duggan, Lisa. *The Twilight of Equality? Neoliberalism, Cultural Politics, and the Attack on Democracy*. Boston: Beacon, 2004.

Dunlap, Riley E., and William R. Carton Jr. "Struggling with Human Exemptionalism: The Rise, Decline and Revitalization of Environmental Sociology." *American Sociologist* 25, no. 1 (1994): 5–30.

Dunlap, Riley E., and K. D. Van Liere. "The 'New Environmental Paradigm': A Proposed Measuring Instrument and Preliminary Results." *Journal of Environmental Education* 9, no. 4 (1978): 10–19.

Dwoskin, Elizabeth. "Why Americans Won't Do Dirty Jobs." *Bloomberg Businessweek*, November 9, 2011. http://www.businessweek.com/magazine/why-americans-wont-do-dirty-jobs-11092011.html#p2.

Eastin, Josh, Reiner Grundmann, and Aseem Prakash. "The Two Limits Debates: 'Limits to Growth' and Climate Change." *Futures* 43, no. 1 (2011): 16–26.

Easton, Robert Olney. *Black Tide: The Santa Barbara Oil Spill and Its Consequences*. New York: Delacorte, 1972.

Ehrenreich, Barbara. *Bright-Sided: How Positive Thinking Is Undermining America*. New York: Metropolitan, 2009.

Ehrlich, Paul. *The Population Bomb*. New York: Ballantine, 1968.

Eisenhower, Dwight D. "Annual Message to the Congress on the State of the Union." January 6, 1955. http://www.eisenhower.archives.gov/all_about _ike/speeches/1955_state_of_the_union.pdf.

Eliasoph, Nina. *Avoiding Politics: How Americans Produce Apathy in Everyday Life*. Cambridge: Cambridge University Press, 1998.

Ellison, Nicole B., Charles Steinfield, and Cliff Lampe. "The Benefits of Facebook 'Friends': Social Capital and College Students' Use of Online Social

Network Sites." *Journal of Computer-Mediated Communication* 12, no. 4 (2007): 1143–68.

"Energy Use (Kg of Oil Equivalent per Capita)." World Bank, n.d. http://data .worldbank.org/indicator/EG.USE.PCAP.KG.OE.

Engdahl, F. William. *A Century of War: Anglo-American Oil Politics and the New World Order*. Ann Arbor, MI: Pluto, 2004.

"Episode #6: Peak Oil Blues." The Extraenvironmentalist, November 11, 2010. http://www.extraenvironmentalist.com/episode-6-peak-oil-blues.

Etzioni, Amitai, ed. *The Essential Communitarian Reader*. Lanham, MD: Rowman & Littlefield, 1998.

Evans, J. P. "Resilience, Ecology and Adaptation in the Experimental City." *Transactions of the Institute of British Geographers* 36, no. 2 (2011): 223–37.

Exxon Mobil Corporation. "Limits to Growth?" *New York Times*, July 25, 2002, A17.

Fagan, Brian M. *The Great Journey: The Peopling of Ancient America*. Gainesville: University Press of Florida, 2003.

Faludi, Susan. *Stiffed: The Betrayal of the American Man*. New York: William Morrow, 1999.

Farber, David, and Jeff Roche. *The Conservative Sixties*. New York: Peter Lang, 2003.

Feil, Ken. *Dying for a Laugh: Disaster Movies and the Camp Imagination*. Middletown, CT: Wesleyan University Press, 2006.

Festinger, Leon. *When Prophecy Fails*. Minneapolis: University of Minnesota Press, 1956.

Finkel, Alexandra. "Nail Polish for Men Is Latest Trend in Grooming." *New York Daily News*, November 3, 2011. http://www.nydailynews.com/life -style/fashion/nail-polish-men-latest-trend-grooming-article-1.971247 #ixzz2UlJms272.

Finucane, Melissa L., Paul Slovic, C. K. Mertz, James Flynn, and Theresa A. Satterfield. "Gender, Race, and Perceived Risk: The 'White Male' Effect." *Health, Risk and Society* 2 (2000): 159–72.

Fisher, Andy. *Radical Ecopsychology: Psychology in the Service of Life*. Albany: State University of New York Press, 2002.

Flynn, William. *Shut Down: A Story of Economic Collapse and Hope*. Charleston, SC: CreateSpace, 2011.

Foner, Eric. *Free Soil, Free Labor, Free Men: The Ideology of the Republican Party Before the Civil War*. New York: Oxford University Press, 1970.

Formisano, Ronald P. *Boston against Busing: Race, Class, and Ethnicity in the 1960s and 1970s*. Chapel Hill: University of North Carolina Press, 1991.

Foster, John Bellamy. "Peak Oil and Energy Imperialism." *Monthly Review* 60, no. 3 (2008): 12–33.

Foster, John Bellamy, and Robert W. McChesney. "The Internet's Unholy Marriage to Capitalism." *Monthly Review: An Independent Socialist Magazine* 62,

no. 10 (2011). http://monthlyreview.org/2011/03/01/the-internets-unholy
-marriage-to-capitalism.

Fox, Justin. "Peak Possibilities." *Time*, November 21, 2007. http://content.time
.com/time/magazine/article/0,9171,1686824,00.html.

Fox-Genovese, Elizabeth. *Feminism without Illusions: A Critique of Individualism*.
Chapel Hill: University of North Carolina Press, 1991.

Franklin, Mark N. "Electoral Participation." In *Controversies in Voting Behavior*,
ed. Richard G. Niemi and Herbert F. Weisberg, 83–99. Washington, DC:
CQ Press, 2001.

Friedrichs, Jörg. "Global Energy Crunch: How Different Parts of the World
Would React to a Peak Oil Scenario." *Energy Policy* 38, no. 8 (2010):
4562–69.

Frum, David. "'Peak Oil' Doomsayers Proved Wrong." CNN Opinion, March 4,
2013. http://www.cnn.com/2013/03/04/opinion/frum-peak-oil.

Funk, McKenzie. *Windfall: The Booming Business of Global Warming*. New York:
Penguin, 2014.

Gabler, Neal. "This Time, the Scene Was Real." *New York Times*, September 16,
2001, sec. 4, p. A2.

Gabriel, Sigmar. "Biodiversity 'Fundamental' to Economics." BBC News,
March 9, 2007. http://news.bbc.co.uk/2/hi/science/nature/6432217.stm.

Garza, Eric L. "The US Energy Information Administration's Faulty Peak Oil
Analysis." *Energy Bulletin*, August 19, 2011. http://www.energybulletin.net/
stories/2011-08-19/us-energy-information-administrations-faulty-peak-oil
-analysis.

Gates, Bill. *The Road Ahead*. New York: Penguin, 1995.

"Gasoline Gangsters Episode 6: Women and Peak Oil." YouTube, February 19,
2011. http://www.youtube.com/watch?v=FQcK-5lE_SE.

Gelderloos, Peter. "An Anarchist Solution to Global Warming." 2010. http://
theanarchistlibrary.org/library/Peter_Gelderloos__An_Anarchist_Solution
_to_Global_Warming.html.

Gent, George. "TV Role in Population Crisis Assayed." *New York Times*,
March 16, 1972, 94.

Ghanta, Praveen. "Is Peak Oil Real? A List of Countries Past Peak." The Oil
Drum, July 18, 2009. http://www.theoildrum.com/node/5576.

Ghosh, Amitav. "Petrofiction: The Oil Encounter and the Novel." *New Republic*,
March 2, 1992, 29–34.

Gibbs, Nancy. "Apocalypse Now." *Time*, July 1, 2002, 38–46.

Gilbey, Ryan. "Climate Change Is Inspiring the Ultimate Scary Movies." *Guardian*, January 1, 2010, 28.

Giroux, Henry A. "Democracy, Patriotism, and Schooling After September 11th:
Critical Citizens or Unthinking Patriots?" In *The Abandoned Generation:
Democracy beyond the Culture of Fear*, 16–45. New York: Macmillan, 2003.

Global Footprint Network. "Data Sources." n.d. http://www.footprintnetwork
.org/en/index.php/GFN/page/data_sources.

———. *Global Footprint Network 2012 Annual Report.* 2012. http://www
.footprintnetwork.org/images/article_uploads/2012_Annual_Report.pdf.

Goldman, Marshall I. *Petrostate: Putin, Power, and the New Russia.* Oxford: Oxford University Press, 2008.

Goldsmith, Jack, and Tim Wu. *Who Controls the Internet? Illusions of a Borderless World.* Oxford: Oxford University Press, 2006.

Goodstein, David. *Out of Gas: The End of the Age of Oil.* New York: Norton, 2005.

Gorelick, Stephen M. *Oil Panic and the Global Crisis: Predictions and Myths.* Hoboken, NJ: Wiley-Blackwell, 2009.

Grafton, Anthony. "History under Attack." *Perspectives on History* 49, no. 1 (2011), http://www.historians.org/publications-and-directories/ perspectives-on-history/january-2011/history-under-attack.

Granovetter, Mark S. "The Strength of Weak Ties." *American Journal of Sociology* 78 (1973): 1360–80.

Grant, Lyle K. "Peak Oil as a Behavioral Problem." *Behavior and Social Issues* 16, no. 1 (2007): 65–88.

Green, Maurice Berkeley. *Eating Oil: Energy Use in Food Production.* Boulder, CO: Westview, 1978.

Greenberg, Paul. "Recipes for Disaster." *New York Times*, April 20, 2008, 27.

Griskevicius, Vladas, Joshua M. Tybur, and Bram Van den Bergh. "Going Green to Be Seen: Status, Reputation, and Conspicuous Conservation." *Journal of Personality and Social Psychology* 98, no. 3 (2010): 392–404.

Grober, Ulrich. "Deep Roots: A Conceptual History of 'Sustainable Development.'" Master's thesis, Wissenschaftszentrum Berlin für Sozialforschung, 2007.

Gronewold, Nathaniel. "Europe's Carbon Emissions Trading—Growing Pains or Wholesale Theft?" *New York Times*, January 31, 2011. http://www .nytimes.com/cwire/2011/01/31/31climatewire-europes-carbon-emissions -trading-growing-pai-74999.html?pagewanted=all.

Gross, Samantha. "Energy Fears Looming, New Survivalists Prepare: Too Late to Save the Planet, They Say, So They Focus on Saving Themselves." MSNBC.com, May 24, 2006. http://www.msnbc.msn.com/id/24808083.

Gupta, Ajay K., and Charles A. S. Hall. "A Review of the Past and Current State of EROI Data." *Sustainability* 3, no. 10 (2011): 1796–1809.

Hackett, Robert A., and Yuezhi Zhao. "Challenging a Master Narrative: Peace Protest and Opinion/Editorial Discourse in the US Press during the Gulf War." *Discourse and Society* 5, no. 4 (1994): 509–41.

Hakes, Jay. "Introduction: A Decidedly Valuable and Dangerous Fuel." *Journal of American History* 99, no. 1 (2012): 19–23.

Hall, Charles A. S. "Unconventional Oil: Tar Sands and Shale Oil—EROI on the Web, Part 3 of 6." The Oil Drum, April 15, 2008. http://www.theoildrum .com/node/3839.

———. "Editorial Introduction to Special Issue on New Studies in EROI (Energy Return on Investment)." *Sustainability* 3, no. 10 (2011): 1773–77.

Hall, C. A. S., C. J. Cleveland, and R. Kaufmann. *Energy and Resource Quality: The Ecology of the Economic Process*. New York: Wiley, 1986.

Hall, Stuart. "Encoding and Decoding in the Television Discourse." Paper presented to the Council of Europe Colloquy, University of Leicester, September 1973.

Hamilton, Clive. *Earthmasters: The Dawn of the Age of Climate Engineering*. New Haven, CT: Yale University Press, 2013.

Hanlon, P., and G. McCartney. "Peak Oil: Will It Be Public Health's Greatest Challenge?" *Public Health* 122, no. 7 (2008): 647–52.

Hardin, Garrett. "Tragedy of the Commons." *Science* 162 (3859): 1243–48.

Hardt, Michael. "Two Faces of Apocalypse: A Letter from Copenhagen." *Polygraph* 22 (2010): 265–74.

"Harris Poll Alienation Index Climbs Again as Two-Thirds of Americans Feel Alienated." Harris Polls, November 12, 2013. http://www.harrisinteractive .com/NewsRoom/HarrisPolls/tabid/447/ctl/ReadCustom%20Default/mid/ 1508/ArticleId/1316/Default.aspx.

Hartley, L. P. *The Go-Between*. London: H. Hamilton, 1953.

Hartman, Thom. *The Last Hours of Ancient Sunlight: The Fate of the World and What We Can Do Before It's Too Late*. New York: Three Rivers, 2004.

Hassan, Robert. "Social Acceleration and the Network Effect: A Defence of Social 'Science Fiction' and Network Determinism." *British Journal of Sociology* 61, no. 2 (2010): 356–74.

Hawks, Tom. "The 50 Points of the Retrosexual/Neosexual Code, for Real Men Today." The Way One Vike Sees It. February 28, 2010. http://onevike .blogspot.com/2010/02/50-points-of-retrosexual-neosexual-code.html.

Hayward, Steven F. *The Age of Reagan: The Conservative Counterrevolution, 1980–1989*. New York: Random House, 2009.

Heinberg, Richard. *The Party's Over: Oil, War, and the Fate of Industrial Societies*. Gabriola Island, BC: New Society, 2003.

———. *Powerdown: Options and Actions for a Post-Carbon World*. West Hoathly: Clairview, 2004.

———. *Peak Everything: Waking Up to a Century of Decline*. Gabriola Island, BC: New Society, 2007.

Helm, Dieter. "Peak Oil and Energy Policy—a Critique." *Oxford Review of Economic Policy* 27, no. 1 (2011): 68–69.

Hemmingsen, Emma. "At the Base of Hubbert's Peak: Grounding the Debate on Petroleum Scarcity." *Geoforum* 41, no. 4 (2010): 531–40.

Hershfield, Hal E., H. Min Bang, and Elke U. Weber. "National Differences in Environmental Concern and Performance Are Predicted by Country Age." *Psychological Science* 25, no. 1 (2014): 152–60.

Hertsgaard, Mark. "Oil Supply Facts, Forecasts Provide Fuel for Thought." *Fort Wayne (IN) Journal-Gazette*, September 11, 2005, 17.

Hill, Gladwin. "Nixon Aide Asks Wide Debate on Desirability of U.S. Growth." *New York Times*, March 30, 1972, 19.

Himmselstein, Jerome L. *To the Right: The Transformation of American Conservatism.* Berkeley and Los Angeles: University of California Press, 1989.

Hitchcock, Peter. "Oil in an American Imaginary." *New Formations* 69, no. 1 (2010): 81–97.

Hixson, Lindsay, Bradford B. Hepler, and Myoung Ouk Kim. "The White Population: 2010." Report no. C2010BR-05. Washington, DC: US Census Bureau, September 2011.

Hoberman, J. "*Nashville* contra *Jaws*; or, The Imagination of Disaster Revisited." In *The Last Great American Picture Show: New Hollywood Cinema in the 1970s*, ed. Alexander Horwath, Thomas Elsaesser, and Noel King, 195–222. Amsterdam: Amsterdam University Press, 2004.

Hobsbawm, Eric. *The Age of Extremes: A History of the World, 1914–1991.* New York: Vintage, 1994.

Hodgson, Godfrey. *America in Our Time: From World War II to Nixon—What Happened and Why.* Garden City, NJ: Doubleday, 1976.

Hofmann, Jeanette. "The Libertarian Origins of Cybercrime: Unintended Side-Effects of a Political Utopia." Discussion Paper no. 62. London: Centre for Analysis of Risk and Regulation, April 2010.

Hoggan, James. *Climate Cover-Up: The Crusade to Deny Global Warming.* Vancouver, BC: Greystone, 2009.

Holzman, David C. "Methane Found in Well Water Near Fracking Sites." *Environmental Health Perspectives* 119, no. 7 (2011): A289.

Höök, Mikael, Robert Hirsch, and Kjell Aleklett. "Giant Oil Field Decline Rates and Their Influence on World Oil Production." *Energy Policy* 37, no. 6 (2009): 2262–72.

Hopkins, Rob. *The Transition Handbook: From Oil Dependency to Local Resilience.* White River Junction, VT: Chelsea Green, 2008.

Howard, James L. "Global Peak Oil Survey 2009." September 27, 2009. http://www.powerswitch.org.uk/portal/index.php?option=com_content&task=view&id=2971&Itemid=77.

Howarth, Robert W., Anthony Ingraffea, and Terry Engelder. "Natural Gas: Should Fracking Stop?" Nature 477, no. 7364 (2011): 271–75.

Howcroft, Debra, Nathalie Mitev, and Melanie Wilson. "What We May Learn from the Social Shaping of Technology Approach." In *Social Theory and Philosophy for Information Systems*, ed. John Mingers and Leslie Willcocks, 329–71. West Sussex: Wiley, 2004.

Huber, Matthew T. "The Use of Gasoline: Value, Oil, and the 'American Way of Life.'" *Antipode* 41, no. 3 (2009): 465–86.

———. *Lifeblood: Oil, Freedom, and the Forces of Capital.* Minneapolis: University of Minnesota Press, 2013.

Hudson, William E. *The Libertarian Illusion: Ideology, Public Policy, and the Assault on the Common Good.* Washington, DC: CQ Press, 2008.

Hulsether, Mark. *Religion, Culture and Politics in the Twentieth-Century United States.* New York: Columbia University Press, 2007.

Humphrey, Kim. *Excess: Anti-Consumerism in the West.* London: Polity, 2009.

Hunter, Alexander. "Editorial: Gore's Bore-a-Thon Day-Long Internet Slide Show Fails to Excite the Planet." *Washington Times*, September 16, 2011. http://www.washingtontimes.com/news/2011/sep/16/gores-bore-a-thon/#ixzz2TlVdELcK.

Hurley, Andrew. *Environmental Inequalities: Class, Race, and Industrial Pollution in Gary, Indiana, 1945–1980.* Chapel Hill: University of North Carolina Press, 1995.

Irwin, Neil. "American Manufacturing Is Coming Back; Manufacturing Jobs Aren't." *Washington Post*, November 19, 2012, http://www.washingtonpost.com/blogs/wonkblog/wp/2012/11/19/american-manufacturing-is-coming-back-manufacturing-jobs-arent.

Jackson, Kenneth T. *Crabgrass Frontier: The Suburbanization of the United States.* New York: Oxford University Press, 1985.

Jacob, Jeffrey. *New Pioneers: The Back-to-the-Land Movement and the Search for a Sustainable Future.* University Park: Pennsylvania State University Press, 1997.

Jacobs, Jane. *The Death and Life of Great American Cities.* New York: Random House, 1961.

Jakob, Doreen. "Crafting Your Way out of the Recession? New Craft Entrepreneurs and the Global Economic Downturn." *Cambridge Journal of Regions, Economy and Society* 6, no. 1 (2013): 127–40.

Jameson, Frederic. "Reification and Utopia in Mass Culture." *Social Text* 1 (1979): 130–48.

Jenkins, Henry. "The Cultural Logic of Media Convergence." *International Journal of Cultural Studies* 7, no. 33 (2004): 33–43.

———. *Convergence Culture: Where Old and New Media Collide.* New York: New York University Press, 2006.

Jenkins, Philip. *Decade of Nightmares: The End of the Sixties and the Making of Eighties America.* New York: Oxford University Press, 2006.

Jeong, Seongeun, Ying-Kuang Hsu, Arlyn E. Andrews, Laura Bianco, Patrick Vaca, James M. Wilczak, and Marc L. Fischer. "A Multitower Measurement Network Estimate of California's Methane Emissions." *Journal of Geophysical Research: Atmospheres* 118, no. 19 (2013): 11–339.

Johnston, Caryl. *After the Crash: An Essay-Novel of the Post Hydrocarbon Age.* Self-published, 2004.

———. "First Peak Oil Novel Published in U.S.!" *Energy Bulletin*, August 15, 2005. http://www.energybulletin.net/node/7921.

Kadaba, Lini S. "The New Retrosexual: Putting the Man Back in Manhood." *Philadelphia Inquirer*, April 23, 2010. http://www.capecodonline.com/apps/pbcs.dll/article?AID=/20100423/LIFE/100429929/-1/LIFE0707.

Kakoudaki, Despina. "Spectacles of History: Race Relations, Melodrama, and the Science Fiction/Disaster Film." *Camera Obscura* 17, no. 2 (2002): 109–53.

Kaminski, Frank. "The Post-Oil Novel: A Celebration!" Seattle Peak Oil Awareness, May 11, 2008. http://www.energybulletin.net/node/44031.

Kasson, John F. *Houdini, Tarzan, and the Perfect Man: The White Male Body and the Challenge of Modernity in America*. New York: Macmillan, 2002.

Kaza, Nikhil, Gerrit-Jan Knaap, Isolde Knaap, and Rebecca Lewis. "Peak Oil, Urban Form, and Public Health: Exploring the Connections." *American Journal of Public Health* 101, no. 9 (2011): 1598–1606.

Keane, Stephen. *Disaster Movies: The Cinema of Catastrophe*. London: Wallflower, 2001.

Kellner, Douglas. *Cinema Wars: Hollywood Film and Politics in the Bush-Cheney Era*. Oxford: Wiley-Blackwell, 2010.

King, Geoff. "'Just Like a Movie?'?: 9/11 and Hollywood Spectacle." In *The Spectacle of the Real: From Hollywood to "Reality" TV and Beyond*, ed. Geoff King, 44–57. Portland, OR: Intellect, 2005.

Kintisch, Eli. *Hack the Planet: Science's Best Hope—or Worst Nightmare—for Averting Climate Catastrophe*. Hoboken, NJ: Wiley, 2011.

Kirby, David, and David Boaz. *The Libertarian Vote in the Age of Obama*. Policy Analysis no. 658. Washington, DC: Cato Institute, January 21, 2010.

Klare, Michael. *Resource Wars: The New Landscape of Global Conflict*. New York: Henry Holt, 2002.

———. *Blood and Oil: The Dangers and Consequences of America's Growing Dependence on Petroleum*. New York: Henry Holt, 2004.

———. *Rising Powers, Shrinking Planet: The New Geopolitics of Energy*. New York: Henry Holt, 2008.

Klatch, Rebecca E. *A Generation Divided: The New Left, the New Right, and the 1960s*. Berkeley and Los Angeles: University of California Press 1999.

Klein, Ezra. "Unpopular Mandate: Why Do Politicians Reverse Their Positions?" *New Yorker*, June 25, 2012, 30–33.

Klinenberg, Eric. *Going Solo: The Extraordinary Rise and Surprising Appeal of Living Alone*. New York: Penguin, 2012.

Kloc, Joe. "America's Invisible Trolley System." *Newsweek*, June 5, 2014, http://www.newsweek.com/2014/06/13/americas-invisible-trolley-system-253455.html.

Kolbert, Elizabeth. *The Sixth Extinction: An Unnatural History*. New York: Henry Holt, 2014.

Kotkin, Joel, "U.S. Will Emerge Stronger: Just As It Did After Downturn of the 1970s." *Atlanta Journal-Constitution*, January 3, 2008, A13.

Kotulak, Ronald. "Fuel, Resources Dwindling: Can America Survive the 20th Century?" *Chicago Tribune*, March 4, 1973, A1.

Krausmann, Fridolin, Karl-Heinz Erb, Simone Gingrich, Helmut Haberl, Alberte Bondeau, Veronika Gaube, Christian Lauk, Christoph Plutzar, and Timothy D. Searchinger. "Global Human Appropriation of Net Primary Production Doubled in the 20th Century." *Proceedings of the National Academy of Sciences* 110, no. 25 (2013): 10324–29.

Kraut, Robert, Michael Patterson, Vicki Lundmark, Sara Kiesler, Tridas Mukophadhya, and William Scherlis. "Internet Paradox: A Social Technology That Reduces Social Involvement and Psychological Well-Being?" *American Psychologist* 53, no. 9 (1998): 1017–31.

Kreijns, Karel, and Paul A. Kirschner. "The Social Affordances of Computer-Supported Collaborative Learning Environments." Paper presented at the ASEE/IEEE Frontiers in Education Conference, Reno, NV, October 2001.

Kruse, Kevin M. *White Flight: Atlanta and the Making of Modern Conservatism.* Princeton, NJ: Princeton University Press, 2005.

Kunstler, James Howard. *The Geography of Nowhere: The Rise and Decline of America's Man-Made Landscape.* New York: Free Press, 1993.

———. *The Long Emergency: Surviving the End of Oil, Climate Change, and Other Converging Catastrophes of the Twenty-First Century.* New York: Atlantic Monthly Press, 2005.

———. "Wake Up, America; We're Driving toward Disaster." *Washington Post,* May 28, 2005, B3.

———. *World Made by Hand.* New York: Atlantic Monthly Press, 2008.

———. *The Witch of Hebron.* New York: Atlantic Monthly Press, 2010.

Kutscher, Ronald E. "Historical Trends, 1950–92, and Current Uncertainties." *Monthly Labor Review,* November 1993, 3–10.

Ladd, Everett. *The Ladd Report.* New York: Free Press, 1999.

Lambert, Jessica G., and Gail P. Lambert. "Predicting the Psychological Response of the American People to Oil Depletion and Declining Energy Return on Investment (EROI)." *Sustainability* 3, no. 11 (2011): 2129–56.

Langer, Gary. "Poll: Public Concern on Warming Gains Intensity." ABC News, March 26, 2006. http://abcnews.go.com/Technology/GlobalWarming/story?id=1750492&page=1.

Lassiter, Matthew D. "Inventing Family Values." In *Rightward Bound: Making America Conservative in the 1970s,* ed. Bruce J. Schulman and Julian E. Zelizer, 13–28. Cambridge, MA: Harvard University Press, 2008.

Lattanzio, Richard K. "Canadian Oil Sands: Life-Cycle Assessments of Greenhouse Gas Emissions." Washington, DC: Congressional Research Service, March 10, 2014. http://fas.org/sgp/crs/misc/R42537.pdf.

Layton, Bradley E. "A Comparison of Energy Densities of Prevalent Energy Sources in Units of Joules per Cubic Meter." *International Journal of Green Energy* 5, no. 6 (2008): 438–55.

Leaning, Jennifer, and Debarati Guha-Sapir. "Natural Disasters, Armed Conflict, and Public Health." *New England Journal of Medicine* 369 (2013): 1836–42.

Lears, T. J. Jackson. *No Place of Grace: Antimodernism and the Transformation of American Culture, 1880–1920.* New York: Pantheon, 1981.

Leblond, Doris. "ASPO Sees Conventional Oil Production Peaking by 2010." *Oil and Gas Journal,* June 30, 2003, 28.

Lee, Martha F. *Earth First! Environmental Apocalypse.* Syracuse, NY: Syracuse University Press, 1995.

Lee, Yueh-Ting, and Martin E. P. Seligman. "Are Americans More Optimistic Than the Chinese?" *Personality and Social Psychology Bulletin* 23 (1997): 32–40.

Leeb, Stephen, and Donna Leeb. *The Oil Factor: Protect Yourself and Profit from the Coming Energy Crisis*. New York: Warner Business, 2005.

Leeb, Stephen, and Glen Strathy. *The Coming Economic Collapse: How You Can Thrive When Oil Costs $200 a Barrel*. New York: Business Plus, 2006.

LeGuin, Ursula K. *The Dispossessed: An Ambiguous Utopia*. New York: Harper & Row, 1974.

Leigh, James. "New Tourism in a New Society Arises from 'Peak Oil.'" *Tourismos: An International Multidisciplinary Journal of Tourism* 6, no. 1 (2011): 165–91.

LeMenager, Stephanie. "The Aesthetics of Petroleum, After Oil!" *American Literary History* 24, no. 1 (2012): 59–86.

———. *Living Oil: Petroleum Culture in the American Century*. Oxford: Oxford University Press, 2014.

Lerner, Max. "Just Imagine! We All Can Avoid a Certain Doomsday." *Los Angeles Times*, March 10, 1972, C9.

Leung, Rebecca. "Woodward Shares War Secrets." CBS News, December 5, 2007. http://www.cbsnews.com/news/woodward-shares-war-secrets.

Leupp, Gary. "The Weekend the World Said No to War: Notes on the Numbers." Counterpunch, February 25, 2003. http://www.counterpunch.org/2003/02/25/the-weekend-the-world-said-no-to-war-notes-on-the-numbers.

Levitt, Tom. "Overfished and Under-Protected: Oceans on the Brink of Catastrophic Collapse." CNN, March 27, 2013. http://www.cnn.com/2013/03/22/world/oceans-overfishing-climate-change.

Lewis, Hyrum. "Historians and the Myth of American Conservatism." *Journal of the Historical Society* 12, no. 1 (2012): 27–45.

Lichterman, Paul. *The Search for Political Community: American Activists Reinventing Commitment*. Cambridge: Cambridge University Press, 1996.

Light, Andrew. "What Is an Ecological Identity?" *Environmental Politics* 9, no. 4 (2000): 59–81.

Lindsay, Hal. *The Late, Great Planet Earth*. New York: Zondervan, 1970.

Lipset, Seymour. *American Exceptionalism: A Double-Edged Sword*. New York: Norton, 1996.

Lipsitz, George. *The Possessive Investment in Whiteness: How White People Profit from Identity Politics*. Philadelphia: Temple University Press, 2006.

Lowenstein, Roger. "What Price Oil?" *New York Times*, October 19, 2008, MM46.

Lowndes, Joseph E. *From the New Deal to the New Right: Race and the Southern Origins of Modern Conservatism*. New Haven, CT: Yale University Press, 2008.

Luce, Henry. "The American Century." *Life*, February 17, 1941, 61.

Lynch, Michael C. "Petroleum Resources Pessimism Debunked in Hubbert Model and Hubbert Modelers' Assessment." *Oil and Gas Journal*, July 14,

2003, 38. http://www.ogj.com/articles/print/volume-101/issue-27/special
-report/petroleum-resources-pessimism-debunked-in-hubbert-model-and
-hubbert-modelers-assessment.html.

———. "'Peak Oil' Is a Waste of Energy." *New York Times*, August 24, 2009, A21.

Maass, Peter. "The Breaking Point." *New York Times*, August 21, 2005, E30.

———. *Crude World: The Violent Twilight of Oil.* New York: Knopf, 2009.

MacKenzie, Donald, and Judy Wajcman, eds. *The Social Shaping of Technology.*
Milton Keynes: Open University Press, 1985.

Majors, Richard. "Cool Pose: Black Masculinity and Sports." In *The Masculinities Reader,* ed. Stephen M. Whitehead and Frank J. Barrett, 209–18.
Cambridge: Polity, 2001.

Malphurs, Ryan. "The Media's Frontier Construction of President George W.
Bush." *Journal of American Culture* 31, no. 2 (2008): 185–201.

Manjikian, Mary. *Apocalypse and Post-Politics: The Romance of the End.* Lanham,
MD: Rowman & Littlefield, 2012.

"Manly Skills." The Art of Manliness, n.d. http://www.artofmanliness.com/
category/manly-skills.

Mann, Charles C. "What If We Never Run Out of Oil?" *Atlantic*, April 24, 2013,
48–63.

Margonelli, Lisa. *Oil on the Brain: Adventures from the Pump to the Pipeline.* New
York: Broadway, 2008.

Marshall, Brent K. "Gender, Race, and Perceived Environmental Risk: The 'White
Male' Effect in Cancer Alley, LA." *Sociological Spectrum* 24 (2000): 453–78.

Martin, Joel W. "Before and Beyond the Sioux Ghost Dance: Native American
Prophetic Movements and the Study of Religion." *Journal of the American
Academy of Religion* 59, no. 4 (1991): 677–701.

Marx, Leo. *The Machine in the Garden: Technology and the Pastoral Ideal in
America.* New York: Oxford University Press, 1964.

Mast, Tom. *Over a Barrel: A Simple Guide to the Oil Shortage.* New York: Hayden,
2005.

Mattson, Kevin. *What the Heck Are You Up To, Mr. President? Jimmy Carter,
America's "Malaise," and the Speech That Should Have Changed the Country.*
New York: Bloomsbury, 2009.

Maugeri, Leonardo. "Squeezing More Oil from the Ground." *Scientific American*
301, no. 4 (2009): 56–63.

May, Elaine Tyler. *Homeward Bound: American Families in the Cold War Era.* New
York: Basic, 1990.

———. *America and the Pill: A History of Promise, Peril, and Liberation.* New York:
Basic, 2010.

McAuliff, Michael. "Torture Like Jack Bauer's Would Be OK, Bubba Says." *New
York Daily News*, October 1, 2007, 22.

McCaffrey, Nancy B., Sara Jane, and Douglas H. Hill. "The Localism Movement: Shared and Emergent Values." *Journal of Environmental Sustainability*
2, no. 2 (2012): 45–57.

McCausland, James. "Scientists' Warnings Unheeded." *Brisbane Courier-Mail*, December 4, 2006. http://www.couriermail.com.au/business/scientists-warnings-unheeded/story-e6freqmx-1111112631991.

McGirr, Lisa. *Suburban Warriors: The Origins of the New American Right*. Princeton, NJ: Princeton University Press, 2001.

McGrath, Ben. "The Dystopians: Bad Times Are Boom Times for Some." *New Yorker*, January 26, 2009, 40–49.

McKay, Brett, and Kate McKay. *The Art of Manliness: Classic Skills and Manners for the Modern Man*. Cincinnati: HOW, 2009.

McKibben, Bill. *Eaarth: Making a Life On a Tough New Planet*. New York: Henry Holt, 2010.

———. "Global Warming's Terrifying New Math." *Rolling Stone*, July 19, 2012, http://www.rollingstone.com/politics/news/global-warmings-terrifying-new-math-20120719.

McKinsey and Co. *Curbing Global Energy-Demand Growth: The Energy Productivity Opportunity*. By Florian Bressand, Diana Farrell, Pedro Haas, Fabrice Morin, Scott Nyquist, Jaana Remes, Sebastian Roemer, Matt Rogers, Jaeson Rosenfeld, and Jonathan Woetzel. n.p.: McKinsey Global Institute, 2007. http://go.nature.com/226nPx.

McMahon, Kathy. "Being Help and Needing Help." Peak Oil Blues, May 12, 2006. http://www.peakoilblues.org/blog/2006/05/12/hello-world.

———. "Do You Have a Panglossian Disorder? or, Economic and Planetary Collapse: Is it a Therapeutic Issue?" Peak Oil Blues, November 13, 2007. http://www.peakoilblues.org/blog/?p=132.

———. "Peak Oil Dating: What Is an 'Ideal PO Mate'?" Peak Oil Blues, August 5, 2008. http://www.peakoilblues.org/blog/?tag=ideal-peak-oil-mate.

———. "Grim Newlywed Sees Scary Future for Those He Loves." Peak Oil Blues, January 14, 2010. http://www.peakoilblues.org/blog/?p=1679.

———. "Stages of Peak Oil Awareness." Peak Oil Blues, September 22, 2010. http://www.peakoilblues.org/blog/?p=2381.

———. *"I Can't Believe You Actually Think That!" A Couple's Guide to Finding Common Ground about Peak Oil, Climate Catastrophe, and Economic Hard Times*. Self-published, 2011.

McNeill, J. R. *Something New under the Sun: An Environmental History of the Twentieth-Century World*. New York: Norton, 2000.

McPherson, Miller, Lynn Smith-Lovin, and Matthew E. Brashears. "Social Isolation in America: Changes in Core Discussion Networks over Two Decades." *American Sociological Review* 71, no. 3 (2006): 353–75.

Meadows, Donella H., Dennis L. Meadows, Jørgen Randers, and William W. Behrens III. *The Limits to Growth*. New York: Universe, 1972.

Mergel, Ines, and Thomas Langenberg. "What Makes Online Ties Sustainable? A Research Design Proposal to Analyze Online Social Networks." Working Paper no. PNG06-002. Cambridge, MA: Program on Networked Governance, John F. Kennedy School of Government, Harvard University, 2006.

Mesch, Gustavo, and Ilan Talmud. "The Quality of Online and Offline Relationships: The Role of Multiplexity and Duration of Social Relationships."
Information Society: An International Journal 22, no. 3 (2006): 137–48.

Messner, Michael A. "The Limits of 'the Male Sex Role': An Analysis of the
Men's Liberation and Men's Rights Movements' Discourse." *Gender and
Society* 12, no. 3 (1998): 255–76.

Miller, Steven P. *Billy Graham and the Rise of the Republican South.* Philadelphia:
University of Pennsylvania Press, 2009.

Miller, Toby. "A Metrosexual Eye on *Queer Guy." GLQ: A Journal of Lesbian and
Gay Studies* 11, no. 1 (2005): 112–17.

Mills, Charles W. *The Racial Contract.* Ithaca, NY: Cornell University Press,
1997.

Mirowski, Philip, and Dieter Plehwe. *The Road from Mont Pèlerin: The Making of
the Neoliberal Thought Collective.* Cambridge, MA: Harvard University Press,
2009.

Mitchel, Dan. "The World According to Kunstler." *New York Times,* July 30,
2005, C5.

Mitchell, Richard J., Jr. *Dancing at Armageddon: Survivalism and Chaos in Modern
Times.* Chicago: University of Chicago Press, 2002.

Mitchell, Timothy. *Carbon Democracy: Political Power in the Age of Oil.* New York:
Verso, 2011.

Mooallem, Jon. "The End Is Near! (Yay!)." *New York Times,* April 16, 2009,
MM28.

Moorhead, James H. *American Apocalypse: Yankee Protestants and the Civil War,
1860–1869.* New Haven, CT: Yale University Press, 1978.

Morozov, Evgeny. *The Net Delusion: The Dark Side of Internet Freedom.* New York:
PublicAffairs, 2011.

———. *To Save Everything, Click Here: The Folly of Technological Solutionism.* New
York: PublicAffairs, 2013.

Motesharrei, Safa, Jorge Rivas, and Eugenia Kalnay. "Human and Nature
Dynamics (HANDY): Modeling Inequality and Use of Resources in the
Collapse or Sustainability of Societies." *Ecological Economics* 101 (2014):
90–102.

Mouawad, Jad. "Oil Innovations Pump New Life into Old Wells." *New York
Times,* March 5, 2007, A1.

Mulligan, Shane. "Energy, Environment, and Security: Critical Links in a Post-
Peak World." *Global Environmental Politics* 10, no. 4 (2010): 79–100.

Murphy, Kevin P. *Political Manhood: Red Bloods, Mollycoddles, and the Politics of
Progressive Era Reform.* New York: Columbia University Press, 2010.

Murtagh, Niamh, Birgitta Gatersleben, and David Uzzell. "Multiple Identities
and Travel Mode Choice for Regular Journeys." *Transportation Research Part
F: Traffic Psychology and Behaviour* 15, no. 55 (2012): 514–24.

Muste, Christopher P. "The Dynamics of Immigration Opinion in the United
States, 1992–2012." *Public Opinion Quarterly* 77, no. 1 (2013): 398–416.

Newman, Peter. "Beyond Peak Oil: Will Our Cities Collapse?" *Journal of Urban Technology* 14, no. 2 (2007): 15–30.

Newport, Frank. "Americans Say Reagan Is the Greatest U.S. President." Gallup, February 18, 2011. http://www.gallup.com/poll/146183/americans-say-reagan-greatest-president.aspx.

Nie, N. H., and L. Erbring. *Internet and Society: A Preliminary Report*. Stanford, CA: Institute for Quantitative Studies of Society, 2000.

Nielsen. "Global Consumers Vote Al Gore, Oprah Winfrey and Kofi Annan Most Influential to Champion Global Warming Cause: Nielsen Survey." July 2, 2007. http://www.eci.ox.ac.uk/news/press-releases/070703pr-climatechamps-world.pdf.

Niemeyer, Philip. "Picturing the Past Ten Years." *New York Times*, December 27, 2009. http://www.nytimes.com/interactive/2009/12/27/opinion/28opchart.html.

Niven, Larry, and Jerry Pournelle. *Lucifer's Hammer*. Santa Monica, CA: Playboy Press, 1977.

Nocera, Joseph. "On Oil Supply, Opinions Aren't Scarce." *New York Times*, September 10, 2005, C1.

Nonnecke, Blair, and Jenny Preece. "Lurker Demographics: Counting the Silent." Paper presented at the Conference on Human Factors in Computing Systems, The Hague, Netherlands, April 2000.

Norgaard, Kari Marie. *Living in Denial: Climate Change, Emotions, and Everyday Life*. Cambridge, MA: MIT Press, 2011.

North, Peter. "Eco-Localisation as a Progressive Response to Peak Oil and Climate Change: A Sympathetic Critique." *Geoforum* 41, no. 4 (2010): 585–94.

Northcott, Michael S. *An Angel Directs the Storm: Apocalyptic Religion and American Empire*. London: SCM, 2007.

"Nostra." "My Movie Influence: Collapse (2009)." Inspired Ground, April 16, 2012. http://www.inspired-ground.com/my-movie-influence-collapse-2009.

Novak, Matt. "Before the Jetsons, Arthur Radebaugh Illustrated the Future." *Smithsonian*, April 2012, 30.

Novick, Peter. *That Noble Dream: The "Objectivity Question" and the American Historical Profession*. Cambridge: Cambridge University Press, 1988.

Nozick, Robert. *Anarchy, State and Utopia*. New York: Basic, 1974.

Olien, Roger M., and Diana Davids Olien. *Oil and Ideology: The Cultural Creation of the American Petroleum Industry*. Chapel Hill: University of North Carolina Press, 2000.

Onion, Rebecca. "Envirogeddon! Is It Time to Start Wishing for the End of the World?" Slate, April 21, 2008. http://www.unz.org/Pub/Slate-2008apr-00291.

Opotow, Susan, and Leah Weiss. "New Ways of Thinking about Environmentalism: Denial and the Process of Moral Exclusion in Environmental Conflict." *Journal of Social Issues* 56, no. 3 (2000): 443–57.

Oreskes, Naomi, and Erik M. Conway. *Merchants of Doubt: How a Handful of Scientists Obscured the Truth on Issues from Tobacco Smoke to Global Warming*. New York: Bloomsbury, 2010.

———. "The Collapse of Western Civilization: A View from the Future." *Daedalus* 142, no. 1 (2013): 40–48.

Owen, Edgar Wesley. *Trek of the Oil Finders: A History of Exploration for Petroleum*. Tulsa, OK: American Association of Petroleum Geologists, 1975.

Owen, Nick A., Oliver R. Inderwildi, and David A. King. "The Status of Conventional World Oil Reserves—Hype or Cause for Concern?" *Energy Policy* 38, no. 8 (2010): 4743–49.

Painter, David S. "Oil and the American Century." *Journal of American History* 99, no. 1 (2012): 24–39.

Parenti, Christian. *Tropic of Chaos: Climate Change and the New Geography of Violence*. New York: Nation, 2011.

Parker, Christopher. "Attitudes on Limits to Liberty, Equality, and Pres. Obama Traits by Tea Party Approval." University of Washington Institute for the Study of Ethnicity, Race and Sexuality, 2010. http://depts.washington.edu/uwiser/Tea%20Party%20Chart%20%5Bpdf%5D-1.pdf.

Pateman, Carol. *The Sexual Contract*. Stanford, CA: Stanford University Press, 1988.

Patterson, James T. *Grand Expectations: The United States, 1945–1974*. New York: Oxford University Press, 1996.

Paxton, Pamela. "Is Social Capital Declining in the United States? A Multiple Indicator Assessment." *American Journal of Sociology* 105, no. 1 (1999): 88–127.

"The Peak Oil Blues." Powering Down: A Journey of Preparation, July 26, 2006. http://poweringdown.blogspot.com/2006/07/peak-oil-blues.html.

"Peak Oil vs. 'Pathological Optimism.'" Ecoshock Radio, October 27, 2010. http://www.ecoshock.info/2010/10/peak-oil-vs-pathological-optimism.html.

Pelletiere, Stephen C. *Iraq and the International Oil System: Why America Went to War in the Gulf*. New York: Praeger, 2001.

Pellow, David N. *Garbage Wars: The Struggle for Environmental Justice in Chicago*. Cambridge, MA: MIT Press, 2004.

Penley, Constance. "Time Travel, Primal Scene, and the Critical Dystopia." In *Close Encounters: Film, Feminism, and Science Fiction*, ed. Constance Penley, Elizabeth Lyon, Lynn Spigel, and Janet Bergstrom, 63–82. Minneapolis: University of Minnesota Press, 1991.

Penny, Sam. *Was a Time When*. Sacramento, CA: TwoPenny, 2012.

Peñuelas, Joseph, and Jofre Carnicer. "Climate Change and Peak Oil: The Urgent Need for a Transition to a Non-Carbon-Emitting Society." *Ambio* 39 (2010): 85–90.

Peterson, Garry. "Visualizing the Great Acceleration—Part II." Resilience Science, December 4, 2008. http://rs.resalliance.org/2008/12/04/visualizing-the-great-acceleration-part-ii.

Pew Research Center for the People and the Press. "Public Sees a Future Full of Promise and Peril." June 22, 2010. http://www.people-press.org/2010/06/22/public-sees-a-future-full-of-promise-and-peril.

———. "Economy Dominates Public's Agenda, Dims Hopes for the Future; Less Optimism about America's Long-Term Prospects." January 20, 2011. http://www.people-press.org/2011/01/20/economy-dominates-publics-agenda-dims-hopes-for-the-future.

———. "Deficit Reduction Rises on Public's Agenda for Obama's Second Term." January 24, 2013. http://www.people-press.org/2013/01/24/deficit-reduction-rises-on-publics-agenda-for-obamas-second-term.

Philippon, Daniel J. "Sustainability and the Humanities: An Extensive Pleasure." *American Literary History* 24, no. 1 (2012): 163–79.

Phillips, Kevin. *American Theocracy: The Peril and Politics of Radical Religion, Oil, and Borrowed Money in the 21st Century.* New York: Viking, 2006.

Phillips-Fein, Kim. "Conservatism: A State of the Field." *Journal of American History* 98, no. 3 (2011): 723–43.

Pierce, William Luther. *The Turner Diaries.* Hillsboro, WV: National Vanguard, 1978.

Piven, Joshua. *The Complete Worst-Case Scenario Survival Handbook.* San Francisco: Chronicle, 2007.

Pollan, Michael. "Why Bother?" *New York Times,* April 20, 2008, MM19.

———. "The Food Movement, Rising." *New York Review of Books,* June 10, 2010, 31–33.

Ponting, Clive. *A Green History of the World: The Environment and the Collapse of Great Civilizations.* New York: Penguin, 1991.

Popke, Jeff. "Latino Migration and Neoliberalism in the U.S. South: Notes toward a Rural Cosmopolitanism." *Southeastern Geographer* 51, no. 2 (2011): 242–59.

"Porn. Peak Oil. Enjoy." http://www.youtube.com/watch?v=vAPf9V3_li0.

Porter, E. D. "Are We Running Out of Oil?" Discussion Paper no. 081. Washington, DC: American Petroleum Institute, 1995.

Porter, Eduardo. "For Insurers, No Doubts on Climate Change." *New York Times,* May 15, 2013, B1.

Postmes, Tom. "The Psychological Dimensions of Collective Action, Online." In *The Oxford Handbook of Internet Psychology,* ed. Adam M. Joinson, 165–86. Oxford: Oxford University Press, 2007.

Potter, David. *People of Plenty: Economic Abundance and the American Character.* Chicago: University of Chicago Press, 1954.

Potter, Sulamith H. "The Cultural Construction of Emotion in Rural Chinese Social Life." *Ethos* 16, no. 2 (1988): 181–208.

Priest, Tyler. "The Dilemmas of Oil Empire." *Journal of American History* 99, no. 1 (2012): 236–51.

Pruit, John C. "Considering Eventfulness as an Explanation for Locally Mediated Peak Oil Narratives." *Qualitative Inquiry* 17, no. 2 (2011): 197–203.

Pushker, Kharecha A., and James E. Hansen. "Implications of 'Peak Oil' for Atmospheric CO2 and Climate." New York: NASA Goddard Institute for Space Studies and Columbia University Earth Institute, n.d. http://arxiv.org/ftp/arxiv/papers/0704/0704.2782.pdf.

Putnam, Robert. *Bowling Alone: The Collapse and Revival of American Community*. New York: Simon & Schuster, 2000.

Quinton, Sophie. "In Manufacturing, Blue-Collar Jobs Need White-Collar Training." *National Journal*, February 27, 2012, http://www.nationaljournal.com/whitehouse/in-manufacturing-blue-collar-jobs-need-white-collar-training-20120226.

Radetzki, Marian. "Peak Oil and Other Threatening Peaks—Chimeras without Substance." *Energy Policy* 38, no. 11 (2010): 6566–69.

Raedle, Joe. "Are We There Yet? Oil Joyride May Be Over." *USA Today*, May 28, 2005, B03.

Ridings, Catherine M., and David Geffer. "Virtual Community Attraction: Why People Hang Out Online." *Journal of Computer-Mediated Communication* 10, no. 1 (2004), http://onlinelibrary.wiley.com/doi/10.1111/j.1083-6101.2004.tb00229.x/abstract.

Roberts, Paul. "Cheap Oil, the Only Oil That Matters, Is Just about Gone." *Harper's*, August 2004, 71–72.

———. *The End of Oil: On the Edge of a Perilous New World*. Boston: Houghton Mifflin, 2004.

Robertson, Thomas. *The Malthusian Moment: Global Population Growth and the Birth of American Environmentalism*. New Brunswick, NJ: Rutgers University Press, 2012.

Robinson, Sally. *Marked Men: White Masculinity in Crisis*. New York: Columbia University Press, 2000.

Roddick, Nick. "Only the Stars Survive: Disaster Movies in the Seventies." In *Performance and Politics in Popular Drama: Aspects of Popular Entertainment in Theatre, Film and Television 1800–1976*, ed. David Bradby, Louis James, and Bernard Sharratt, 243–70. Cambridge: Cambridge University Press, 1980.

Rodgers, Daniel. *The Age of Fracture*. Cambridge, MA: Harvard University Press, 2011.

Rogers, Amy. *Petroplague*. New York: Diversion, 2011.

Romano, Andrew, and Tony Dokoupil. "Men's Lib." Newsweek, September 27, 2010, 43.

Romano, Renee C. "Not Dead Yet: My Identity Crisis as a Historian of the Recent Past." In *Doing Recent History: On Privacy, Copyright, Video Games, Institutional Review Boards, Activist Scholarship, and History That Talks Back*, ed. Claire Bond Potter and Renee C. Romano, 23–44. Athens: University of Georgia Press, 2012.

Rome, Adam. "'Give Earth a Chance': The Environmental Movement and the Sixties." *Journal of American History* 90, no. 2 (2003): 525–54.

Rosenthal, Elisabeth, and Andrew W. Lehren. "Profits on Carbon Credits Drive Output of a Harmful Gas." *New York Times*, August 8, 2012, A1.

Rosin, Hanna. "Who Wears the Pants in This Economy?" *New York Times*, September 16, 2012, MM8.

Roszak, Theodore, Mary E. Gomes, and Allen D. Kanner. *Ecopsychology: Restoring the Earth, Healing the Mind*. San Francisco: Sierra Club, 1995.

Roth, Michael S. "Foucault's 'History of the Present.'" *History and Theory* 20, no. 1 (1981): 32–46.

Ruppert, Michael. "Colin Campbell on Oil: Perhaps the World's Foremost Expert on Oil and the Oil Business Confirms the Ever More Apparent Reality of the Post–9-11 World." From the Wilderness, October 23, 2002. http://www.fromthewilderness.com/free/ww3/102302_campbell.html.

———. *Crossing the Rubicon: The Decline of the American Empire at the End of the Age of Oil*. New York: New Society, 2004.

———. *Confronting Collapse: The Crisis of Energy and Money in a Post Peak Oil World*. White River Junction, VT: Chelsea Green, 2009.

Rymph, Catherine E. *Republican Women: Feminism and Conservatism from Suffrage to the New Right*. Chapel Hill: University of North Carolina Press, 2006.

Saad, Lydia. "Conservatives Remain the Largest Ideological Group in U.S." Gallup, January 12, 2012. http://www.gallup.com/poll/152021/conservatives-remain-largest-ideological-group.aspx.

Sabin, Paul. *Crude Politics: The California Oil Market, 1900–1940*. Berkeley and Los Angeles: University of California Press, 2004.

Sack, Kevin. "Apocalyptic Theology Revived by Attacks." *New York Times*, November 23, 2001, A17.

Salam, Reihan. "Heralding the End Times." *New York Sun*, March 5, 2008, Arts and Letters section, 15.

Sampson, Anthony. *The Seven Sisters: The Great Oil Companies and the World They Shaped*. New York: Bantam, 1975.

Samuel, Lawrence R. *Future: A Recent History*. Austin: University of Texas Press, 2009.

Saunders, Megan I., Javier Leon, Stuart R. Phinn, David P. Callaghan, Katherine R. O'Brien, Chris M. Roelfsema, Catherine E. Lovelock, Mitchell B. Lyons, and Peter J. Mumby. "Coastal Retreat and Improved Water Quality Mitigate Losses of Seagrass from Sea Level Rise." *Global Change Biology* 19, no. 8 (2013): 2569–83.

Scarrow, Alex. *Last Light*. London: Orion, 2007.

———. *Afterlight*. London: Orion, 2010.

Schreiber, Ronnee. *Righting Feminism: Conservative Women and American Politics*. New York: Oxford University Press, 2008.

Schumacher, E. F. *Small Is Beautiful: A Study of Economics as If People Mattered*. London: Blond & Briggs, 1973.

Schwartz, Brian S., Cindy L. Parker, Jeremy Hess, and Howard Frumkin. "Public Health and Medicine in an Age of Energy Scarcity: The Case of Petroleum." *American Journal of Public Health* 101, no. 9 (2011): 1560–67.

Scott, David L. *Wall Street Words: An A to Z Guide to Investment Terms for Today's Investor.* Boston: Houghton Mifflin, 2010.

Scott, James C. *Weapons of the Weak: Everyday Forms of Peasant Resistance.* New Haven, CT: Yale University Press, 1985.

———. *Domination and the Arts of Resistance: Hidden Transcripts.* New Haven, CT: Yale University Press, 1990.

Scranton, Roy. "Learning How to Die in the Anthropocene." *New York Times,* November 10, 2013. http://opinionator.blogs.nytimes.com/2013/11/10/learning-how-to-die-in-the-anthropocene.

Self, Robert O. *American Babylon: Race and the Struggle for Postwar Oakland.* Princeton, NJ: Princeton University Press, 2003.

SeokWoo, Son, Neil F. Tandon, Lorenzo M. Polvani, and Darryn W. Waugh. "Ozone Hole and Southern Hemisphere Climate Change." *Geophysical Research Letters* 36, no. 15 (2009): L15705.

Serwer, Adam, Jaeah Lee, and Zaineb Mohammed. "Now That's What I Call Gerrymandering!" *Mother Jones,* November 14, 2012, http://www.motherjones.com/politics/2012/11/republicans-gerrymandering-house-representatives-election-chart.

Shah, Sonia. *Crude: The Story of Oil.* Toronto: Seven Stories, 2004.

Shahani, Nishant. *Queer Retrosexualities: The Politics of Reparative Return.* Plymouth: Lehigh University Press, 2012.

Shierholz, Heidi. "Immigration and Wages—Methodological Advancements Confirm Modest Gains for Native Workers." Briefing Paper no. 255. Washington, DC: Economic Policy Institute, February 4, 2010.

Shils, Edward. "Ideology: The Concept and Function of Ideology." In *The Encyclopedia of Philosophy* (8 vols.), ed. Paul Edwards, 4:974–78. New York: MacMillan, 1967.

Silver, Nate. "Poll Finds a Shift toward More Libertarian Views." *New York Times,* June 20, 2011. http://fivethirtyeight.blogs.nytimes.com/2011/06/20/poll-finds-a-shift-toward-more-libertarian-views.

Simmons, Matthew R. *Twilight in the Desert: The Coming Saudi Oil Shock and the World Economy.* Hoboken, NJ: Wiley, 2005.

Skrimshire, Stefan, ed. *Future Ethics: Climate Change and Apocalyptic Imagination.* London: Continuum, 2010.

Slotkin, Richard. *Gunfighter Nation: The Myth of the Frontier in Twentieth-Century America.* New York: HarperPerennial, 1993.

Slusser, George. "Pocket Apocalyse: American Survivalist Fictions from *Walden* to *The Incredible Shrinking Man.*" In *Imagining Apocalypse: Studies in Cultural Crisis,* ed. David Seed, 118–35. New York: St. Martin's, 2000.

Smil, Vaclav. "Peak Oil: A Catastrophic Cult and Complex Realities." *World Watch Magazine* 19, no. 1 (2006): 22–24.

Smith, Amanda. "The Transition Town Network: A Review of Current Evolutions and Renaissance." *Social Movement Studies* 10, no. 1 (2011): 99–105.

Smith, Henry Nash. *Virgin Land: The American West as Symbol and Myth.* Cambridge, MA: Harvard University Press, 1950.

Smith, Marc A. "Invisible Crowds in Cyberspace: Mapping the Social Structure of the Usenet." In *Communities in Cyberspace*, ed. Marc A. Smith and Peter Kollock, 195–218. New York: Routledge, 1999.

"SnakelaD." "The Movie That Changed My Life." Amazon.com, n.d.. http://www.amazon.com/Collapse-Michael-Ruppert/product-reviews/B003CJXJ8Q?pageNumber=7 (post now removed).

Solberg, Carl. *Oil Power.* New York: Signet, 1976.

Somashekhar, Sandhya, and Peyton Craighill. "Slim Majority Back Gay Marriage, *Post*-ABC Poll Says." *Washington Post*, March 18, 2011. http://articles.washingtonpost.com/2011-03-18/politics/35207422_1_gay-marriage-pro-gay-marriage-group-anti-gay-marriage.

Song, Felicia Wu. *Virtual Communities: Bowling Alone, Online Together.* New York: Peter Lang, 2009.

Sparks, Glenn G. *Media Effects Research.* Boston: Wadsworth, 2010.

Speth, James Gustave. *The Bridge at the End of the World: Capitalism, the Environment, and Crossing from Crisis to Sustainability* . Ann Arbor, MI: Caravan, 2008.

Spicer, Dave. *Deadly Freedom.* New York: Cambridge House, 2008.

Spitz, Peter H. *Petrochemicals: The Rise of an Industry.* New York: Wiley, 1988.

Srnicek, Nick. "Assemblage Theory, Complexity and Contentious Politics: The Political Ontology of Gilles Deleuze." Master's thesis, University of Western Ontario, 2007.

Stabile, Carol A. *White Victims, Black Villains: Gender, Race and Crime News in US Culture.* New York: Routledge, 2006.

Stanley, Bruce. "Oil Experts Draw Fire for Warning." *Dubuque (IA) Telegraph Herald*, May 27, 2002, B5.

Steffen, William L., Angelina Sanderson, Peter Tyson, Jill Jäger, Pamela A. Matson, Berrien Moore III, and Frank Oldfield. *Global Change and the Earth System: A Planet under Pressure.* Berlin: Springer, 2005.

Steingraber, Sandra. *Living Downstream: An Ecologist's Personal Investigation of Cancer and the Environment.* 1997. 2nd ed. Philadelphia: Da Capo, 2010.

Stern, Nicholas. *Review on the Economics of Climate Change.* London: HM Treasury, 2006.

Stine, Jeffrey K. "Natural Resources and Environmental Policy." In *The Reagan Presidency: Pragmatic Conservatism and Its Legacies*, ed. W. Elliott Brownlee and Hugh Davis Graham, 233–58. Lawrence: University Press of Kansas, 2003.

Stott, Philip. "Hot Air + Flawed Science = Dangerous Emissions." *Wall Street Journal*, April 2, 2001, A22.

Strahan, David. *The Last Oil Shock: A Survival Guide to the Imminent Extinction of Petroleum Man.* New York: John Murray, 2007.

Streeter, Thomas. *The Net Effect: Romanticism, Capitalism, and the Internet.* New York: New York University Press, 2010.

Stretesky, Paul B., Sheila Huss, Michael J. Lynch, Sammy Zahran, and Bob Childs. "The Founding of Environmental Justice Organizations across U.S. Counties during the 1990s and 2000s: Civil Rights and Environmental Cross-Movement Effects." *Social Problems* 58, no. 3 (2011): 330–60.

Strub, Whitney. "Further into the Right: The Ever-Expanding Historiography of the U.S. New Right." *Journal of Social History* 42, no. 1 (2008): 183–94.

Sugrue, Thomas J. *The Origins of the Urban Crisis: Race and Inequality in Postwar Detroit.* Princeton, NJ: Princeton University Press, 1996.

Sullivan, J. Courtney, and Courtney E. Martin, eds. *Click: When We Knew We Were Feminists.* New York: Seal, 2010.

Svetkey, Benjamin. "Lava Is a Many Splendored Thing." *Entertainment Weekly,* August 25, 1997, 3.

Szeman, Imre. "System Failure: Oil, Futurity, and the Anticipation of Disaster." *South Atlantic Quarterly* 106, no. 4 (2007): 805–23.

———. "The Cultural Politics of Oil: On Lessons of Darkness and Black Sea Files." *Polygraph* 22 (2010): 33–45.

Szeman, Imre, Jennifer Wenzel, and Patricia Yaeger, eds. *Fueling Culture: Energy, History, Politics.* Bronx, NY: Fordham University Press, 2015.

Tainter, Joseph. *The Collapse of Complex Societies.* Cambridge: Cambridge University Press, 1988.

Tarbell, Ida M. *The History of the Standard Oil Company.* New York: McClure, Phillips, 1904.

Taylor, Bron. "Exploring Religion, Nature and Culture." *Journal for the Study of Religion, Nature and Culture* 1, no. 1 (2007): 5–24.

Taylor, Dorceta E. *Race, Class, Gender, and American Environmentalism.* Washington, DC: US Department of Agriculture, 2002.

Teles, Steven, and Daniel A. Kenney. "Spreading the Word: The Diffusion of American Conservatism in Europe and Beyond." In *Growing Apart? America and Europe in the Twenty-First Century,* ed. Jeffrey Kopstein and Sven Steinmo, 136–69. New York: Cambridge University Press, 2008.

Tertzakian, Peter. *A Thousand Barrels a Second: The Coming Oil Break Point and the Challenges Facing an Energy Dependent World.* New York: McGraw-Hill, 2006.

Thomashow, Mitchell. *Ecological Identity: Becoming a Reflective Environmentalist.* Cambridge, MA: MIT Press, 1996.

Tinkle, Lon. *Mr. De: A Biography of Everette Lee DeGolyer.* New York: Little, Brown, 1970.

"Towns Banking Their Own Currency." BBC News, April 2, 2008. http://news .bbc.co.uk/2/hi/uk_news/wales/south_west/7326212.stm.

Townsend, Robert B. "History under the Hammer." *Perspectives on History* 49, no. 1 (2011): 1286–1306.

Traister, Aaron. "'Retrosexuals': The Latest Lame Male Catchphrase." *Salon*, April 7, 2010. http://www.salon.com/2010/04/07/retrosexuals_silliness.

Tranter, Paul, and Scott Sharpe. "Children and Peak Oil: An Opportunity in Crisis." *International Journal of Children's Rights* 15, no. 1 (2007): 181–97.

———. "Escaping Monstropolis: Child-Friendly Cities, Peak Oil and Monsters, Inc." *Children's Geographies* 6, no. 3 (2008): 295–308.

Truman, Harry S. "Annual Message to the Congress on the State of the Union" (January 5, 1949). In *Public Papers of the Presidents of the United States: Harry S. Truman*, 1–7. Washington, DC: US Government Printing Office, 1964.

Tubella, Ima. "Television and Internet in the Construction of Identity." In *The Network Society: From Knowledge to Policy*, ed. Manuel Castells, 257–68. Washington, DC: Johns Hopkins Center for Transatlantic Relations, 2008.

Tuck, Stephen. "The New American Histories." *Historical Journal* 48, no. 3 (2005): 811–32.

Turkle, Sherry. *Life on the Screen: Identity in the Age of the Internet*. New York: Simon & Schuster, 1997.

———. *Alone Together: Why We Expect More from Technology and Less from Each Other*. New York: Basic, 2011.

Turner, Fred. *From Counterculture to Cyberculture: Stewart Brand, the Whole Earth Network, and the Rise of Digital Utopianism*. Chicago: University of Chicago Press, 2006.

Turner, Jack. *Awakening to Race: Individualism and Social Consciousness in America*. Chicago: University of Chicago Press, 2012.

Turner, Frederick Jackson. *The Significance of the Frontier in American History*. Great Ideas. Harmondsworth: Penguin, 2008.

Tuveson, Ernest Lee. *Redeemer Nation: The Idea of America's Millennial Role*. Chicago: University of Chicago Press, 1968.

Urstadt, Bryan. "Imagine There's No Oil: Scenes from a Liberal Apocalypse." *Harper's*, August 2006, 31–40.

US Department of Labor. US Bureau of Labor Statistics. "Highlights of Women's Earnings in 2011." Report 1038. October 2012. http://www.bls.gov/cps/cpswom2011.pdf.

US Energy Information Administration. "Gasoline and Diesel Fuel Update," n.d. http://www.eia.gov/petroleum/gasdiesel.

———. "U.S. Refinery Yield." n.d. http://www.eia.gov/dnav/pet/PET_PNP_PCT_DC_NUS_PCT_A.htm.

———. Independent Statistics and Analysis. "International Petroleum (Oil) Consumption Tables (Thousand Barrels per Day)." n.d. http://www.eia.gov/emeu/international/oilconsumption.html.

US PIRG Education Fund. *Transportation in Transition*. December 2013. http://www.uspirg.org/sites/pirg/files/reports/US_Transp_trans_scrn.pdf.

Vance, Erik. "Letter from the Sea of Cortez: Emptying the World's Aquarium; The Dismal Future of the Global Fishery." *Harper's*, August 2013, 53–62.

Vasterman, Peter L. M. "Media-Hype: Self-Reinforcing News Waves, Journalistic Standards and the Construction of Social Problems." *European Journal of Communication* 20, no. 4 (2005): 508–30.

Vidal, John. "Global Warming Causes 300,000 Deaths a Year, Says Kofi Annan Thinktank." *Guardian*, May 29, 2009. http://www.theguardian.com/environment/2009/may/29/1.

Vietor, Richard H. K. *Energy Policy in America since 1945: A Study of Business-Government Relations*. New York: Cambridge University Press, 1984.

"Voters Issue Strong Rebuke of Incumbents in Congress." Gallup, April 7, 2010. http://www.gallup.com/poll/127241/voters-issue-strong-rebuke-incumbents-congress.aspx.

Wackernagel, Mathis, and William Rees. *Our Ecological Footprint: Reducing Human Impact on the Earth*. Gabriola Island, BC: New Society, 1996.

Watcharasukarn, Montira, Shannon Page, and Susan Krumdieck. "Virtual Reality Simulation Game Approach to Investigate Transport Adaptive Capacity for Peak Oil Planning." *Transportation Research* 46, no. 2 (2012): 348–67.

Weber, Timothy. *Living in the Shadow of the Second Coming: American Premillennialism, 1875–1982*. New York: Oxford University Press, 1979.

Weinstein, Neil D. "Optimistic Biases about Personal Risks." *Science* 246, no. 4935 (1989): 1232–33.

Weisman, Steven R. "Reagan Tiptoes around Some Economic Liabilities." *New York Times*, September 26, 1982, sec. 4, p. 4.

Wellman, Barry. "Little Boxes, Glocalization, and Networked Individualism." In *Digital Cities II: Computational and Sociological Approaches*, ed. Mokato Tanabe, Peter van den Besselaar, and Toru Ishida, 10–25. Berlin: Springer, 2002.

Western, John W. *Selling Intervention and War: The Presidency, the Media, and the American Public*. Baltimore: Johns Hopkins University Press, 2005.

Whitehead, Stephen M., and Frank J. Barrett, eds. *The Masculinities Reader*. Cambridge: Polity, 2001.

Whyte, William H. *The Organization Man*. New York: Simon & Schuster, 1956.

Wiegley, Russell Frank. *The American Way of War: A History of United States Military and Policy*. New York: Macmillan, 1973.

Wilentz, Sean. *The Age of Reagan: A History, 1974–2008*. New York: HarperCollins, 2008.

Williams, Alex. "Duck and Cover: It's the New Survivalism." *New York Times*, April 6, 2008, ST-1.

Williamson, Harold F., Ralph L. Andreano, and Arnold R. Daum. *The American Petroleum Industry: The Age of Energy, 1899–1959*. Evanston, IL: Northwestern University Press, 1963.

Williamson, Vanessa, Theda Skocpol, and John Coggin. *The Tea Party and the Remaking of Republican Conservatism*. New York: Oxford University Press, 2012.

Willson, Robert. "*Jaws* as Submarine Movie." *Jump Cut: A Review of Contemporary Media* 15 (1977): 32–33.

Wilson, Brian. "Ethnography, the Internet, and Youth Culture: Strategies for Examining Social Resistance and 'Online-Offline' Relationships." *Canadian Journal of Education* 29, no. 1 (2006): 307–28.

Winch, Peter, and Rebecca Stepnitz. "Peak Oil and Health in Low- and Middle-Income Countries: Impacts and Potential Responses." *American Journal of Public Health* 101, no. 9 (2011): 1225–34.

Winn, J. Emmett. "Mad Max, Reaganism and the Road Warrior." *Kinema: A Journal for Film and Audiovisual Media* 23 (2009): 57–75.

Winner, Langdon. "Technology as Forms of Life." In *Readings in the Philosophy of Technology*, ed. David M. Kaplan, 103–13. Oxford: Rowman & Littlefield, 2004.

Wojcik, Daniel. *The End of the World as We Know It: Faith, Fatalism and Apocalypse in America*. New York: New York University Press, 1999.

Wood, John H., Gary R. Long, and David F. Morehouse. "World Conventional Oil Supply Expected to Peak in 21st Century." Offshore, April 1, 2003. http://www.offshore-mag.com/articles/print/volume-63/issue-4/technology/world-conventional-oil-supply-expected-to-peak-in-21st -century.html.

Woodward, Bob. "Greenspan Is Critical of Bush in Memoir: Former Fed Chairman Has Praise for Clinton." *Washington Post*, September 16, 2007, A1.

Woodward, C. Vann. "The Age of Reinterpretation." *American Historical Review* 66, no. 1 (1960): 1–19.

Woodward, Llewellyn. "The Study of Contemporary History." *Journal of Contemporary History* 1, no. 1 (1966): 1–13.

World Commission on Environment and Development. *Our Common Future*. Oxford: Oxford University Press, 1987.

Worster, Donald. *Nature's Economy: A History of Ecological Ideas*. New York: Cambridge University Press, 1985.

———. *The Wealth of Nature: Environmental History and the Ecological Imagination*. New York: Oxford University Press, 1993.

Wright, Kevin B. "Researching Internet-Based Populations: Advantages and Disadvantages of Online Survey Research, Online Questionnaire Authoring Software Packages, and Web Survey Services." *Journal of Computer-Mediated Communication* 10, no. 3 (2005), http://onlinelibrary.wiley.com/journal/10.1111/(ISSN)1083-6101.

WWF International. *Living Planet Report 2012: Biodiversity, Biocapacity and Better Choices*. 2012. http://awsassets.panda.org/downloads/1_lpr_2012_online _full_size_single_pages_final_120516.pdf.

Yardley, Jonathan. "All Built Out of Ticky-Tacky." *Washington Post*, May 30, 1993, X03.

Yergin, Daniel. *The Prize: The Epic Quest for Oil, Money and Power*. New York: ABC, 1996.

———. "Imagining a $7-a-Gallon Future." *New York Times*, April 4, 2004, WK1.

———. "There Will Be Oil." *Wall Street Journal*, September 17, 2011, C1.

"Y2K Millennium Bug—Wave 3." Gallup (poll sponsored by USA Today/ National Science Foundation), August 25–29, 1999. http://brain.gallup .com/documents/questionnaire.aspx?STUDY=P9908039&p=2.

Yuen, Eddie. "The Politics of Failure Have Failed: The Environmental Movement and Catastrophism." In *Catastrophism: The Apocalyptic Politics of Collapse and Rebirth*, by Sasha Lilley, David McNally, Eddie Yuen, and James Davis, 15–43. Oakland, CA: PM, 2012.

Zaretsky, Natasha. *No Direction Home: The American Family and the Fear of National Decline, 1968–1980*. Chapel Hill: University of North Carolina Press, 2007.

Zelizer, Julian. "Reflections: Rethinking the History of American Conservatism." *Reviews in American History* 38, no. 2 (2010): 367–92.

Zelizer, Julian E., and Bruce J. Schulman, eds. *Rightward Bound: Making America Conservative in the 1970s*. Cambridge, MA: Harvard University Press, 2008.

Zernike, Kate, and Megan Thee-Brenan. "Discontent's Demography: Who Backs the Tea Party?" *New York Times*, April 15, 2010, A1.

Zhao, Lin, Lianyong Feng, and Charles A. S. Hall. "Is Peakoilism Coming?" *Energy Policy* 37, no. 6 (2009): 2136–38.

Ziser, Michael. "Home Again: Peak Oil, Climate Change, and the Aesthetics of Transition." In *Environmental Criticism for the Twenty-First Century*, ed. Stephanie LeMenager, Teresa Shewry, and Ken Hiltner, 181–95. New York: Routledge, 2011.

Ziser, Michael, and Julie Sze. "Climate Change, Environmental Aesthetics, and Global Environmental Justice Cultural Studies." *Discourse* 29, nos. 2/3 (2007): 384–410.

Zogby International. "Americans Link Katrina, Global Warming." 2006. http:// www.zogby.com/News/ReadNews.dbm?ID=1161 (now defunct).

Index

biosphere, 10, 108, 156
birth control pill, 62
birth rate. *See* population
Black, Brian C., 63
blogs, 3, 40, 48, 52, 92, 130, 132, 141; influential, 13, 18, 33–35, 37, 40, 95, 125
blood for oil, 13, 52–53, 192n26, 197n100
Boggs, Carl, 87
Boulianne, Shelley, 92
Bowling Alone (Putnam), 5, 175n15
Bright-Sided (Ehrenreich), 44, 225n47
Brundtland Commisssion. *See* World Commission on Environment and Development
Buell, Frederick, 70, 154, 157
Bush administration (George W.), 7, 13, 52–53, 86, 140, 189n7, 193n44; rise of conservatism and, 7, 49, 86, 105–6
business as usual, 21, 44, 158, 182n15

Campbell, Colin, 47–48, 50, 53, 98
Can, Kris, 37, 95
Candide (Voltaire), 42, 187n94
capitalism: American imperialism and, 15, 23, 25, 72, 104, 124–25; anticapitalism and, 15, 123–25; collapse of, 21, 27, 68, 104, 121, 124–25; future of, 20, 27, 121–22; growth-oriented, 10, 25, 68, 70, 83, 127, 196n81; libertarianism and, 82–84, 91
carbon sinks, 10, 20, 73, 75, 156
Carnegie, Dale, 44
Carson, Rachel, 66–67
Carter, Jimmy (American president), 69–71, 189n113
Cassandra, 117. *See also* Oily Cassandra
Castells, Manuel, 93
Cato Institute, 84, 199n18
Cerulo, Karen A., 13, 44
Chavez, Leo R., 143
Cheneyism, 13, 42. *See also* Bush administration
Chicken Littles, 39, 70
Christian evangelism, 83, 105–6, 137, 148, 209n32
civic engagement. *See* depoliticization; network effect; public discourse
climate change, 5, 16, 150–51, 156–57; as anthropogenic, 10, 13, 47, 57, 74, 108, 128, 222n17; climate change denial, 4, 44, 125, 159–61

Clinton, Bill (American president), 71–72, 86, 91, 140
Club of Rome (Meadows, Meadows, Randers, Behrens), 68, 70, 72
coal, 19, 64–65
Cobb, Kurt, 116, 130, 181n2
Cold War, 34, 53, 61–62, 67, 106–9, 185n55
Cold War masculinity, 138
Collapse (Diamond), 21
Collapse of Complex Societies (Tainter), 21
collective action, 20, 71, 78–81, 86, 159, 162–63, 206n105; identity and, 3–4, 94–95, 97–102, 109, 127–28; traditional, 7, 14, 24–25, 29, 61, 68, 87, 185n55
Collins, Robert M., 60, 64. *See also* growthmanship
Committee on Growth, 61
commodity chains, 11–12
communism, 67, 82–83, 195n74
communitarian ideals, 3, 16, 81, 84, 97, 104, 199n10
Community Supported Agriculture (CSA), 28
Competitive Enterprise Institute, 84
conservativism, 3, 6–8, 52, 68, 83, 118, 200n21, 200n24; antienvironmentalism and, 70–73, 160, 194n54; big government and, 82, 86, 162, 201n25; post-apocalyptic culture and, 123, 134–36, 143, 148, 202n47. *See also* libertarianism; Reagan; Tea Party
conspiracy theories, 6, 39, 49, 184n38
consumption patterns, 10–11, 14, 23, 59, 65–66, 120, 211n51; ecological footprint of, 28, 69, 73, 156–59, 224n37; of hyperconsumption, 25, 54–55, 60–62, 116, 150, 155, 178n46; limits of, 75, 127, 151, 156–58, 224n28; peakists and, 26–27, 33, 41, 43, 49, 153
convergence culture, 48, 192n23
Coontz, Stephanie, 142, 201n26
Crabgrass Frontier (Jackson), 34
crowds, 14, 98–99, 101
CSA (Community Supported Agriculture), 28
cyberlibertarianism, 3, 15, 76, 89–92, 101, 154, 161
cyberspace. *See* Internet

dark age, 21, 36
Darwin, Charles, 66